M000316398

"Gene's long-awaited follo
Not An Option is a checkl
greatness. *Tough and Competent* is much more than a history
of U.S. manned spaceflight—it is a graduate-level text on goal
setting, team building, risk mitigation and acceptance, self-
discipline, and introspection. There is no organization I can think
of, be it civilian, government service, or the military, that would
not benefit from Gene's latest book."

Lieutenant Colonel (Retired) Mitch Utterback,
U.S. Army Special Forces

"For innovators surfing the edge of chaos, *Tough and Competent*
is your mentoring guide to become a 21st century leader, building
intrepid teams that harness risks and change the World."

Captain Richard Champion de Crespigny AM,
Pilot in Command, Qantas flight QF32

"While the world knows Gene Kranz as the guy in Mission
Control with the vest, I know him as a man who has a unique
ability to lead a team of extraordinary people dealing with
extremely complex operations that have consequences that.
range from historic to disastrous."

"Dutch" von Ehrenfried, Mission Controller,
Aviation and Space Author

"Team Chemistry, well described in Kranz's book, *Tough and
Competent* is the handbook I use to develop the technical, personal
and trusting relationships within my team that elevates and sustains
operations performance. Kranz's life story also has a personal
connection. At American Airlines I am currently in the same
position as his Air Force jet instructor and mentor Wendell Dobbs."

Capt. John P. Dudley
Managing Director Flight Operations American Airlines

"Kranz's *Tough And Competent* is a fascinating follow-up to *Failure Is Not An Option*, taking us way deeper into the decision-making processes that, famously, saved the Apollo 13 crew - and which we learn later helped save the Skylab mission, too. The secret sauce, we discover, is the unique team chemistry that Kranz forged in Mission Control after the Apollo 1 fire, in which expertise, innovation and an uncompromising team spirit combined, time and again, to safely confront risk. Essential reading."

Paul Marks, Spaceflight Journalist,
Aerospace America and New Scientist

"A masterclass in leadership and teamwork from THE legendary NASA Mission Controller Gene Kranz, *Tough and Competent* gives a valuable insight into how his dictums for effective management and cool reactions in a crisis, helped put humans into space, land on the Moon and, most importantly, brought them home safely. A must read in how close-knit teams develop 'IT' – an almost unquantifiable ability to thrive under pressure."

Tim Robinson,
Editor in Chief, AEROSPACE magazine
Royal Aeronautical Society

"The core lesson of *Tough and Competent* is that leadership and culture, not software or hardware, are the essential ingredients in managing unimaginable complexity. Gene Kranz continues to be an inspiration for generations of leaders daring to push us all forward and build resilience in a fragile world."

Jon Ostrower,
Editor-in-chief, The Air Current

"Success in spaceflight relies on leadership, and Flight Director Gene Kranz is the leader whose teams put Apollo 11 on the Moon, rescued the crippled Apollo 13, recovered from the Apollo fire, saved Skylab, and ushered the shuttle into service. "Tough and Competent"--Gene's indispensable story of a lifetime spent leading and succeeding—should be on America's reading list, and yours."

Tom Jones, Veteran Shuttle Astronaut,
Scientist, Pilot, and Author of *Space Shuttle Stories*

"In my role at the National Aviation Hall of Fame, I have close personal relationships with many of the most impactful people of aviation and aerospace. There are two groups of legends, those who seek to ensure that future generations are reminded of their legacy and those who continue to serve humanity without consideration of their legacy. This second group, who humbly accept their place in history and continue to engage until they are gone, are the selfless aerospace legends who share themselves not for glory, but out of service. In *Tough and Competent*, Gene Kranz shows why he is in this latter group and how his leadership is authentic, relatable, and adaptable. Leaders like Gene are not made, they are grown, and this book intimately provides a flight plan for generations to come."

Amy Spowart,
President and CEO of the National Aviation Hall of Fame

"I enjoyed firsthand experience working with the mission control teams as a Capcom and backup crewmember on three lunar missions. In his book *Tough and Competent* Kranz captures the essence of the development of the leadership and team chemistry, the IT! of the Control teams who worked around the clock to overcome a seemingly impossible challenge to provide a great Hollywood ending to our mission."

Fred Haise,
Astronaut Apollo 13

FLIGHT DIRECT

Tough and Competent

LEADERSHIP AND TEAM CHEMISTRY

EUGENE F. KRANZ

FORMER NASA DIRECTOR OF
MISSION OPERATIONS

gatekeeper press™
Tampa, Florida

Tough and Competent: Leadership and Team Chemistry

Published by Gatekeeper Press
7853 Gunn Hwy., Suite 209
Tampa, FL 33626
www.GatekeeperPress.com

All photos courtesy of NASA
Cover Photo: NASA Gemini 9 Press Conference, May 9
2007 RNASA painting by Pat Rawlings

Library of Congress Control Number: 2022947967

ISBN (hardcover): 9781662933301
ISBN (paperback): 9781662933318
eISBN: 9781662933325

"Ultimately, leadership is not about glorious crowning acts. It's about keeping your team focused on a goal and motivated to do their best to achieve it, especially when the stakes are high, and the consequences really matter. It is about laying the groundwork for others' success, and then standing back and letting them shine."

—Astronaut Chris Hadfield

CONTENTS

The Moon and Beyond

Challenge

Foundations of Mission Control

Introduction—Captain Richard Champion de Crespigny AM

Where were you when you heard about 9/11? I remember flying an Airbus A380, north of Darwin on our route from Hong Kong to Sydney. Where were you when Neil Armstrong first stepped on the moon? I remember being in the school assembly hall watching the black and white live broadcast. 9/11 was shocking. The moon landing was extraordinary. Both significant emotional events changed the world and our lives. Both are etched clearly into my mind.

The brain commits significant emotional events to long-term memory. We happily recall good events. Horrific events create trauma that can persist to our last breath.

The three significant emotional events that significantly advanced technology in the 20th century were two world wars and NASA's Space Program. The wars allocated unlimited resources to improve aeronautics, electronics, and nuclear industries. Whilst we benefitted from these new technologies, no one wants war as the catalyst for radical change.

Memories of NASA's Apollo project were different, a bold and exciting mission to land a man on the Moon and return him safely to the Earth before the end of the 1960s. Although it was a politically motivated venture to assert presence in space, the politics ended there. Unlike warring nations, NASA was open in its missions, transparent in its delivery, and invited the world to participate. Anyone in the world could follow and share the progress of successes and failures of the Mercury, Gemini, and Apollo programs. I am Australian and although we did not own or fund NASA, we followed NASA's mission to the moon as though it were our own.

NASA's journey to the moon was a phenomenally expensive venture that changed the world but the return on investment was priceless. Inertial Navigation, Fly-By-Wire controls, rockets, trajectories, computers, communications—things we take for granted today, were all developed during missions. Skylab, the Shuttle, and the International Space Station are all derivatives from Apollo that continued discovery in space.

Positive disruptors in energy, water, and transportation require national funding and exceptional leadership. To achieve Earth-changing projects like these, funding, leadership, and teamwork at the scale of NASA's Space Program is imperative.

Humanity needs another Apollo-type program to stimulate cohesion, creativity, adaption, and success. A mission that every person wants, that creates something amazing and unites the World. To achieve the extraordinary, we don't just want resilient people who can survive a crisis, we need that special breed of human that defies the sceptics, rejects the status quo, rejects the impostor syndrome, seeks out the seemingly impossible task, then makes it happen. We need to bring out the best in humanity, for humanity.

Today's world leaders are not as fearless as President Kennedy in 1961 when he announced the USA's Space Effort. The result is that humanity is not as resilient as we would wish it to be. In fact, humanity is brittle.

Our mission must be to seek out the greatest challenges on Earth—projects that frighten us—with advanced technology that requires massive human participation and risk. We must take humanity where it has never been before, because the hardest challenges unite and advance us. In summary, to help the Earth we need leadership, passion, discipline, competence, confidence, responsibility, teamwork, toughness and grit.

Gene's first book, *Failure is not an Option* is a chronicle of the Mercury, Gemini, and Apollo missions. It relates what happened during those remarkable times that changed the world.

But this book left me questioning the HOWs and WHYs. Why did Kennedy announce his intent to put a man on the moon and back within just nine years, when there were almost no skills, technology, and structures in place to do it? How much did it cost? What was Gene's leadership style and how did he develop it? How did he build and lead multiple extraordinarily skilled teams with an average age of just twenty-seven years? How could Gene "ad lib" his prophetic words that became known as the "Kranz Dictum" that changed the culture in NASA operations after the Apollo 1 fire? And why did Gene Kranz exceed his authority to make that speech when I think it was Chris Kraft's responsibility to deliver it?

Tough and Competent is the logical successor to Gene's first book. It explains Gene's upbringing which set his values and beliefs. These are his WHYs, that in turn determined his HOWs, that in turn determined his WHATs. When you read *Tough and Competent* you'll understand the man, his roots, and the people that mentored him. You will understand why he is humble yet sharp, fiercely determined, and with an incredible intellect. When you finally get the full spectrum of Kranz the leader, then you will understand his thoughts and anticipate his actions and words. And when you grasp the gravitas of Kranz leading teams of extraordinary people that accomplished the impossible, then you will understand the critical reasons for NASA's success.

I've discovered that most people in life seek boundaries and don't want to rock the boat. Yet the explorers, inventors and doers that we respect, are people that reject the status quo, question everything, trust their judgement, regularly break rules, and take on any challenge, regardless of their orders. These leaders are eminent mentors that we would gladly give our lives to support, and their values are the ingredients for every person to reach their full potential to have a happy, meaningful, and respected life.

Tough and Competent is the leadership and teamwork book for the 21st century:

If you want to step into the arena to be not just a great leader but a World leader—then this is the book for you.

If you want to understand the human factors (the chemistry, or *IT!*) to build and maintain extraordinary, dynamic, and resilient teams of experts, all sharing one common mental model of values, situational awareness, roles, tasks and trust, and acting as one— then this is the book for you.

If you are fearful of taking risks and want to feel "bulletproof—not gun shy", or courageous, because you are confident that whatever happens, that you have the procedures to do, or that you can create novel solutions to do, the things required to survive—then this is the book for you.

Finally, if you have a project to improve planet Earth, a project as visionary as NASA's Apollo project, which needs the Right Stuff in leadership and the Right Stuff in teamwork, a project that will create happy significant emotional memories that humanity will never forget, then this is your guide.

Captain Richard Champion de Crespigny AM
Pilot in Command, Qantas flight QF32

Quotes

For World-changing innovators, Kranz shows us the characteristics we need to be visionary and intrepid leaders and the chemistry in tenacious and resilient teams to achieve the seemingly impossible.

For innovators surfing the edge of chaos, *Tough and Competent* is your mentoring guide to become a 21st century leader, building intrepid teams that harness risks and change the World.

Tough and Competent should be mandatory reading at every political and CEO school.

Tough and Competent—the Elements of World changing Leadership and Teamwork

The secret to building and leading intrepid and resilient teams that change the World.

Qantas Flight 32
November 4, 2010

Australian Richard de Crespigny was pilot in command of an A380 superjumbo, with 469 people onboard when one of the four engines exploded Energised shards of metal sliced through the wing and fuselage, cutting 650 wires, leaving gaping holes, damaging control systems, puncturing fuel tanks, and destroying half the networks. Twenty-one of the twenty-two systems were affected.

The captain and crew worked through a confusing cascade of over 100 computer messages and distracting alarms from the on-board computers. Two hours later with one engine out, severely overweight, and with limited control, a landing was accomplished. With fuel leaking from the aircraft, white hot brakes, and unable to shut down an engine, de Crespigny kept passengers onboard for two hours until it was safe to disembark with no injuries.

Leadership and team chemistry were key elements in QF32's success. De Crespigny and his team ignored faulty checklists, created novel solutions and used undocumented procedures to survive this Black Swan event.

In the 2016 Australia Day Honours, Richard de Crespigny was made a "Member of the Order of Australia" (AM) for his significant service to the aviation industry, both nationally and internationally, particularly to flight safety and to the community.

PROLOGUE

THE ARENA

It is not the critic who counts; nor the man who points out how the strong man stumbles, or where the doer of deeds could have done them better. The credit belongs to the man who is actually in the arena.
—*President Theodore Roosevelt,*
Citizenship in a Republic

I was present in Mission Control for over 100 launches and lived by two simple principles: "Treat every mission as if it was the first" and "Do what is best for the crews who fly." Mission Control is a living organism, a complex structure of interdependent and subordinate elements whose relations and properties are largely determined by their function during the course of a mission. We are the first responders in spaceflight, accountable for all actions consistent with crew safety and mission success.

Our work in manned spaceflight involves enormous energy, complex systems, operations in a vacuum, velocity measured in miles/second, null gravity, extreme temperatures and on occasion split-second irreversible actions. Charles Murray, describing my work as Flight Director in his book, *Apollo: The Race to the Moon,* states that,

It isn't just the authority and visibility that set the Flight Director apart, but the job itself. . . . Know in technical detail one of the most complex machines ever made. Master a complex flight plan and a huge body of mission rules. Piece together tiny and often unconnected bits of information from multiple sources coming to you at the same time. Do this under the gaze of the world in situations that might give you only seconds to make life and death decisions. It was a job for not just anybody, it was a job that had no equal.[1]

It is here that I again became familiar with death; indeed, it was an element of the lifework I accepted and in the course of my career I lost a number of good friends to tragic accidents. But I believe it is all part of living a life "in the arena." Being in the arena means taking on risk and it means failing from time to time. But there is no achievement without risk. And there is no success without failure.

Many different images have been referenced to describe Mission Control. Chris Kraft, the conceptual architect, described it as a "cathedral," and for many years I referred to it as the "leadership laboratory" that assembled the next generation for space flight. But I believe "arena," as Teddy Roosevelt used it in a different time, is the best term we can use, Mission Control as the place where individually and as a team we confronted the unexpected, sought to control risk, and reached to go further as explorers. "In the Arena we know the triumph of high achievements," Roosevelt wrote, "and at the worst if we fail, at least fail while daring greatly. So that our place shall never be with the cold and timid souls who know neither victory nor defeat."

***

With the completion of the Gemini 12 Mission in November 1966, I moved without pause to support the pending Apollo 1 launch in late January 1967. The intensity of the effort was not new, it was the way we had lived during Gemini. We were used to surrendering weekends and holidays. I received my first Flight Director assignment during Project Gemini.

I now had nine missions under my belt and, along with Chris Kraft and John Hodge, was selected for the Apollo 1 flight test. I had worked with Gus Grissom on early missions and did the planning for Ed White's Extra Vehicular Activity (EVA). Roger Chaffee was the new guy, a Navy pilot who during the Cuban Missile Crisis took pictures of the missile complexes.

Shortly after midnight on January 27, 1967, I was again seated at Mission Control's Flight Director console in Houston, Texas, laboriously updating the launch pad test procedures with the last-minute changes sent over from Cape Canaveral in Florida (the Cape). I was tired from the previous day's "Plugs-In Test" with the backup crew, Wally Schirra, Donn Eisele, and Walt Cunningham. After a brief break from the prior day's test, I decided to remain throughout the early morning with ground controllers and the Cape communications team to sort out the communications problems we experienced in the previous day's test.

Today's "Plugs-Out Test" was a full rehearsal for the launch day countdown. Test power was applied about 7:00 a.m. CST and the various spacecraft systems were activated. Except for some procedural changes the test continued normally until close to noon when I called Kraft, and we began our team handover for the crew insertion. With the countdown proceeding, I returned to the office. Throughout the afternoon, the Mission Control and Cape teams were plagued by a variety of problems, mostly communications and an unhappy crew. A hold was called at 6:27 p.m. CDT to prepare for power transfer. Four minutes later, fire exploded in the spacecraft and in a pure-oxygen environment the temperature reached about one thousand degrees. The three crew members, Grissom, White, and Chaffee, died in seventeen seconds. The command module ruptured. All voice and data communications terminated three seconds after the rupture. Flames and gasses flowed from the rupture and ignited other combustibles. In this inferno, attempts were made to remove the hatches but they only opened inwards and the heat from the fire made opening impossible.

I was at home. LM Branch chief Jim Hannigan, a neighbor ran to our house and yelled, "There was a fire on the pad, and they think the crew is dead." I literally hopped in my car and tore off to Mission Control. Kraft had cut off all outside communications and locked the entrances. Finding the entrance locked, I went around to the freight elevator and up to the second floor.

There was a deathly silence in the control room when I entered. Hodge was the first to speak to me. "It was gruesome," he said, clutching his pipe in his teeth as he tried to remain focused. Hodge knew death, he had been a boy during the London blitz and a flight test engineer.

Kraft finished talking to the surgeon, walked over, and said, "Deke thinks we were lucky. When the spacecraft ruptured there was molten metal dribbling out."

John Llewellyn, a tough Marine who fought at the Chosin Reservoir in the Korean War was without words. He and Dutch von Ehrenfried, controllers with whom I played judo, were crying.

After a horrible hour, one-by-one the controllers left after securing the data, logbooks, and the video and voice tapes with the word "fire." We locked the control room doors and gravitated to the "Singing Wheel," our watering hole, to begin our wake. The proprietor, Nelson Bland, cleared the building except for the controllers. Beer was on the house and soon worried wives began showing up trying to take husbands home. The night was one of subdued, limited conversation. Controllers would start to talk, then shudder and break into tears. We mourned our crew and whatever naïveté we had left.

I had been through this before when we lost pilots. We would drink and sing our sad fighter pilot songs; we had no wives to take us home. Singing and drinking was our outlet.

There were no songs in the "Singing Wheel" that night and when we went home, we were drained of emotion, changed in ways we could physically feel but not describe.

The weekend felt infinite, the soul searching intense and very personal. The television, newspapers, and calls from reporters across the nation amplified the pain. I don't know how the reporters got my phone number, but I finally unplugged the phone. Aviators have only two fears: fear of failure and fear of fire. I relived the crashes, the fires, and the times I saw men die. My only thought was how and why we failed the Apollo 1 crew.

Highly graphic news came Saturday morning from Jules Bergman, the ABC space reporter. The test purpose, countdown, entering the hold, the power transfer followed by the flash in the cabin video camera, and the moment of death were described. With a cold wind blowing the flags at half-mast, at 2:00 a.m. the bodies were removed. After Sunday mass I called Hodge, and we scheduled a briefing in the Building 30 auditorium for Monday.

Monday, January 30, 1967, followed a sleepless night. Hodge finished speaking, laying out the accident investigation team and schedule. As I climbed the stairs to the stage, the crews' words, "fire in the cockpit," echoed in my mind and soured my gut. When I turned and looked at the sea of faces before me, I could feel their grief and their anger that someone had screwed up. Those seated before me, while Gemini veterans, were mostly young. We worked daily to control the risks inherent in our work and were well prepared to handle it in orbit, but this happened on the launch pad, and we had no way to help.

Death was no stranger to me or to the few controllers who saw service in Korea. Most of the controllers had now experienced a friend's death for the first time and now were acutely aware of the risks in our work. I was their leader, angry because someone screwed up, and our crew died.

Harry Carroll, my flight test boss, taught me that words *"light the fire."* They express the most powerful force available to humanity. We can choose to use this force constructively with words of encouragement or destructively using words of despair. Words can heal or humble. Mine were intended to create an unstoppable drive. I had to get my controllers out of the shock and establish the mindset, *"never again."* Resilience, unflagging determination, and conviction must now propel us toward our lunar goal. Whatever might happen in our gut, we could not nor cannot waver or lose sight of our cause.

On stage, the words came to me. At Holloman I had a card on my desk that showed a biplane crashed high into a tree with the caption, "Aviation unlike the sea, is terribly intolerant of carelessness, incapacity, or neglect." I began speaking slowly, deliberately, and defined accountability.

Space flight will never tolerate carelessness, incapacity, and neglect. Somewhere, somehow, we screwed up. It could have been in design, build, or test. Whatever it was, we should have caught it. We were too gung-ho about the schedule, and we locked out all the problems we saw each day in our work. Every element of the program was in trouble and so were we. The simulators were not working, Mission Control was behind in virtually every area, and the flight and test procedures changed daily. Nothing we did had any shelf life. Not one of us stood up and said, "Dammit stop!" I do not know what Thompson's committee will find as the cause, but I know what I find. *We are the cause! We were not ready! We did not do our job!* We were rolling the dice, hoping that things would come together by launch day, when in our hearts, we knew it would take a miracle. We were pushing the schedule and betting that the Cape would slip before we did. I then gave my instructions for how we would respond to this tragedy.

From this day forward, flight control will be known by two words: "Tough" and "Competent." TOUGH means we are forever accountable for what we do, or what we fail to do. We will never again compromise our responsibilities. Every time we walk into Mission Control, we will know what we stand for. COMPETENT means we will never take anything for granted. We will never be found short in our knowledge and in our skills. Mission Control will be perfect. When you leave this meeting today, you will go to your office and the first thing you will do there is to write "Tough" and "Competent" on your blackboards. It will never be erased. Each day when you enter the room, this phrase will remind you of the price paid by Grissom, White, and Chaffee. These words are the price of admission to the ranks of Mission Control.

Astronauts Grissom, White, Chaffee

My controllers left the auditorium in silence, changed forever. "Tough and Competent" entered our common vocabulary and my address became known as "The Kranz Dictum." That day we developed the mindset and the individual fire to go to the moon. The day we landed on the moon we would fulfill the pledge we made to Gus, Ed, and Roger.

BEGINNING

1

FOUNDATION

No matter how bleak or menacing a situation may appear, it does not entirely own us. It can't take away our freedom to respond, our power to take action
—Ryder Carrol

I was born in 1933 during the Great Depression in Toledo, Ohio. My father, a World War I veteran, died when I was seven years old and like many others of his time, he had no life insurance. One of my greatest losses is that I never got to know my father. Of my few recollections is him listening nightly to the radio as Edward Murrow reported on the German Blitz and the terror bombing of London.

Toledo was a military transit hub, so after my father's death, my mother moved my two sisters and I to a home in West Toledo. We lived near the American Legion Hall, so she turned the home into a military boarding house to pay the bills. My mother was a strong German, Catholic, and a Republican. She set an example in all things and was tough because she needed to be tough. She had high expectations for her three children and sacrificed to provide them opportunities. Her gifts to her children were integrity, belief in themselves, and the drive to work for their dreams. My sisters were in their late teens. Louise had entered nursing school and Helen would soon follow.

My father, Leo Kranz, a WWI Veteran; mother, Margaret Kranz; and me, Eugene Kranz

Our street was mostly Polish-Irish, much like the Norman Rockwell's paintings of American culture that graced the *Saturday Evening Post* cover. Our neighbor, John McHugh, was an Irish cop, who whistled as he walked home from the streetcar in his crisp, blue uniform swinging a nightstick. The nuns from the nearby Saint Agnes School reported discipline problems to McHugh and, one by one, the kids got a little talking to. He was the surrogate father for the boys on the street whose fathers were fighting in the war. The Willys-Overland Jeep plant was at the end of our street and, by the time we were in the fifth and sixth grade, one of the rites of passage was to take a Jeep from the storage lot, drive it between the rows of Jeeps to the test track, and down the hill to Ten Mile Creek. Another was to stand next to the railroad tracks at a curve and see who flinched first and stepped back as a steam locomotive passed. All in all, it was a fun place to grow up and, while many of us were bruised, none were broken.

I was a paperboy and, every morning and evening, as I rolled papers, I read the headlines. "Extra" editions were exciting; walking the streets late at night or during the day, I would shout, "Extra! Extra! Read all about it." Windows and doors opened, and people asked, "What happened?" I gave them the news: "Doolittle Raid on Tokyo" or "Americans Sink Japanese Fleet." Like the TV commentators today, we were part of American history, the first people to pass the news, and we felt important. My work as a paperboy taught me responsibility like making sure that newspapers were delivered on time to every customer regardless of weather and sickness. Keeping my bicycle working was simply a responsibility to accomplish my job. This simple task to deliver a product to a customer would put me on a path that prepared me for more demanding roles in my life.

My mother crowded the family into a single bedroom, so we could take three to five boarders in the two remaining bedrooms, providing them two meals, and washing their clothes. In the evenings, the GIs talked of the great land, sea, and air battles, bringing the war close to our home. Louise, my oldest sister was blonde, lively, and always focused on a far horizon. When home from nursing school, she played songs of the 40's on the piano while listening to the radio, occasionally improvising with a bit of boogie woogie. Helen was the flirt, a brunette with soft brown eyes, a hearty laugh, and a high school junior, which was the age of many of our soldiers. Mom lost two of her "boarder sons" and we hung a Gold Star in the window for the ones who never returned. After the war "Punchy" Grimes, a Navy middleweight boxer, and Reinhart Brandt, an infantryman with Patton's Army, returned and worked in the Sunoco refineries. The men going through our home were the brothers I never had, and they taught me about risk, sacrifice, and country. With fewer boarders after the war, times got tough and we lived on potatoes, root vegetables, small game, and fish provided by an uncle. Sunday mornings were special, fried cornmeal mush and honey were for breakfast.

I carried two jobs during high school to support my mother. My high school teachers established a class schedule to accommodate morning paper deliveries and as an evening stock boy at the A&P Grocery. Grocery stores were not open on Sundays and the teachers provided the next weeks' assignments on Friday so I could complete assignments over the weekend. My science, math, and history teachers at Central Catholic High School discovered I had math and science abilities and began coaching me for the entrance examination for the Naval Academy. With their help, I obtained a Naval Reserve Officers Training Corps (NROTC) scholarship to the

University of Notre Dame, Indiana, and a Naval Academy appointment. I reported to the Naval Reserve Station in Detroit, Michigan, for the physical exam in August 1950 and weeks later began my senior year at Central Catholic. Shortly after the school year began, I received a telegram indicating that I had failed the physical examination and was scheduled for a glucose retest in Detroit. Weeks later, I received a telegram that I was medically disqualified for appointment to the Naval Academy and to Notre Dame as well. My dream died—I lost my chance to become a naval aviator and go to college.

When I showed the telegram to Sister Mark, she brushed back tears, lifted her chin, and said, "This is not the end; we will find the way." Sister Mark and my mother canvassed scholarship sources, finally finding an Ohio Elks Association $500 loan for children of deceased veterans of World War I. That was a start. Sister Mark released me from my last period for the rest of the year for a few more work hours every week. My mother and sister Louise chipped in, and we sold my father's stamp albums. With every resource on the table, I could start college in August 1951. We never owned a car and hitchhiking provided my ability to move beyond the range of my bike. When the war ended the National Air Races provided the proving ground and a showcase for the new products in Cleveland and Detroit. I had been hitching long distance going to the airshows so hitching 475 miles to Parks College in East Saint Louis, Illinois, was not much of a challenge.

Aviation pulp magazines of the 50s were filled with advertisements for many of the small flying schools, among them were Spartan, Embry-Riddle, and Parks Air College which was founded in 1927. I selected Parks because money was the issue. With only a $500 loan it offered a trimester program and I could graduate in three years. Parks with crossed cinder runways, and Stearman PT-17 trainers droning over the classrooms became my place of dreams. Dormitories were WWII barracks, many classrooms were in hangars, and the shops were used to maintain the airplanes for pilot training during the war. Ten percent of all pilots who flew in World War II received their instruction from one of Oliver Parks' flying schools. America's top living ace, Francis S. Gabreski, was one of 27,000 pilots who received their primary flight training at Parks. Parks was a School of Saint Louis University and classes went year around except for July and two weeks at Christmas. New students were assigned older students as roommates, and I was double bunked with three Korean War Navy veterans who provided counseling and a jolt if I got behind in studies. The campus closed at 10:00 p.m. and any curfew violation brought mandatory work time. Each student was required to spend two hundred hours work time on campus polishing floors, working in the mess hall, or on the flight line to reduce college cost.

In 1954 the Parks student population consisted of 150 military and 350 civilians with curriculums in engineering, meteorology, aircraft maintenance, and flight training. The engineering curriculum consisted of 16-18 credit hours per semester with several non-credit evening courses. The engineering courses were augmented by hands-on aircraft maintenance courses as electives. In a time before computers, several courses were done at night on Frieden Electromechanical Calculators. I enjoyed the Air Science (AFROTC) courses and the evening drill periods, which helped pay for tuition. The third year at Parks went rapidly and the studies got tougher. My final engineering course GPA was 2.45 and, while low by today's standards, it came with extensive hands-on knowledge of the technologies of the day. There is nothing better than working as a team, rebuilding a large rotary aircraft engine, installing it in a test stand, and seeing our work settle into the roar of a successful test run.

The fifty-two-member graduation class consisted of Korean veterans, six Israelis, and thirty-eight civilians like myself. The Scott Air Force Base Band played during our commissioning and after the Oath of Office, the gold bar was pinned on my epaulets. The next step would be Air Force wings and I prayed I was ready. My mother, sister Louise, and Uncle Albert were with me that happy day. They had pooled their resources and bought me a green, 1954 Plymouth Belvedere coupe as a graduation present. I had flown an airplane, but I had yet to drive a car.

2

RENAISSANCE MAN

I have been impressed with the urgency of doing. Knowing is not enough; we must apply. Being willing is not enough; we must do.

—Leonardo da Vinci

Reflecting back on the day I finished college and the years that led up to it, I think of the many people whose wisdom, caring, and most of all, their belief in me that made it all possible. It is they who prepared me for "My Time," the new life I was about to begin.

McDonnell Aircraft Company (MAC) was a great friend of Parks College and by hiring engineering and maintenance graduates often provided the employment "bridge" between college and the Air Force. I was assigned to pilot training class 56M with a reporting date of March 1955. I applied to MAC, and was happy when they provided an opportunity to work through my hiatus before I reported for duty. In the 1950s, graduates entering the aircraft industry could work in drafting or data reduction. I chose data reduction because it was not a desk job and was attached to the flight test department, a huge bullpen housing several hundred engineers. Pilots were in a glass-enclosed area and flight test engineers, instrumentation, and data reduction personnel were situated overlooking the aircraft ramp.

McDonnell was where I met an early mentor. It was my first day on the job when a gruff voice called out behind me, "Are you Kranz? Where have you been?"

I turned around and there he was: a balding, cherubic bundle of energy gripped at the neck by a bow tie and held together by colorful suspenders that stood out like a lighthouse on an ebony night. I was a little surprised to be challenged like that from the start but, truth be told, I had no idea what I was in for. In the end, Harry Carroll was a leader to emulate, perhaps the best I would ever know.

He moved out quickly and I followed him to a desk covered with rolls of paper. He shoved them against the wall, pushed me into the chair, and said, "When you reduce these oscillograph rolls and learn to read the data, you will know more about what happened during a flight test than the pilot, flight test engineer, and the designer. These rolls of paper are like novels. It is up to you to sense the plot, get the meaning, and determine whether flight objectives were satisfied. You must watch to see if we are getting too close to the edge." Breathlessly, he concluded, "This is the best job in flight test!" Then he stepped back, chuckled. and said, "Get started."

My first job was to reduce the data from the flight tests of the Convertiplane, a cross between a helicopter and a plane. The rotor blades were driven by pulse jet on the rotor blade tips. After each flight my task was to measure every squiggle of data, and provide a cumulative "damage estimate" to the designers. The oscillograph rolls contained everything Carroll said they would and each day I learned a new lesson. Like most great leaders, Harry taught by example, assigned responsibility early, and set a standard for excellence. He communicated a zest for work in the quest for understanding every nuance of the flight test data that the team was responsible to analyze.

One day, I left the recorders running after the preflight checks on a test. The light beam oscillograph (LBO) recorder paper ran out shortly after takeoff and, with no data recorded, the test was a bust and had to be redone. His message to me was that "mistakes are how we learn." Assuming accountability for my error, his message to senior management was that he, Harry, was to blame. "I've got your back," was a critical lesson in bringing on new personnel, assessing performance, and building confidence.

You didn't work *for* Harry. You did your work as it fit into the larger scheme of the flight test program. By the end of the third month, I was reducing the "quick look" data and between flights, I sat with the test pilot and flight test engineer, reviewing pilot's notes, and validating flight test objectives. Owning the work forced me to make decisions, verifying that the flight objectives were satisfied, whether any test points were missed, and whether instrumentation changes were needed. This was heady stuff for a recent college graduate and after I started walking the flight line with the instrumentation engineers supporting pre- and post-flight activities, I was on top of the world. The only better job was in the cockpit and I could feel the clock counting down to the start of my flight training.

From Harry, I learned the "Flight Test Mentality." Many flights appeared similar and the work tedious, but he taught me that every flight should be treated as a "first flight." Later in my career, when I was asked to critique the 2003 *Columbia* space shuttle accident where all seven crew members perished, I concluded that we had lost our "Flight Test Mentality."

* * *

Born in Chicago on December 10, 1918, Harry was from my father's generation. A true "Renaissance Man," Harry was a World War II bomber pilot, inventor, engineer, poet, dinner theatre actor, and Eagle Scout leader. During the war, he flew the B-17 Flying Fortress for the 15th Air Force, 99th Bomb Group out of Foggia, Italy. In between eleven missions targeting German oil reserves near Ploiesti, Romania, he would fit in a trip to the opera. A date with Madame Butterfly one night; the cratering of Nazi targets the next. Harry's first combat mission with the "Diamondbacks," as the 99th was called, was April 6, 1944, with Ploiesti as the target. This mission was so hazardous that if they returned crewmembers were given credit for two missions flown. Later, Harry switched to the 8th Air Force, 381st Bomb Group out of Ridgewell, England, completing twenty-six missions, many of them in support of the Normandy invasion. All of this was, to say the least, difficult work. In fact, if you ever think you are not capable of "going on," consult the war diaries of a World War II bomber group and read about those who made it through or the many who were lost. Of the aircraft assigned to Harry's 381st Bomb Group from 1943-1944, 80 percent were lost or salvaged: *four out of five.*

The twenty-six missions flown by Harry's crew through September 1944 included Berlin, Kiel, Hamburg, Munich, and Schweinfurt as well as missions supporting the Allied advance through Holland and France. On an early mission over Hamm, Germany, targeting an important railway marshalling yard, two engines were shot out and he was unable to return to his base, so he landed at an emergency airstrip at Manston, Kent, England.

On a later mission, approaching Berlin in a "dog of an airplane," he struggled to remain in formation, becoming a target for German fighters. "It was a battle royal, lasting about fifteen minutes," he told me. "My tail gunner was shooting at a guy coming in at six or seven o'clock and the flight engineer started shooting at the same guy, saying he got him, but I was too busy with other things at the time. I looked at number four, which had a big hole, was leaking oil and fuel, and just a little fire. I feathered and hit the extinguisher, and I had other things to do. I told the bombardier, 'Jettison the bombs.'" Trailing smoke, Harry found cloud cover and the arrival of escort fighters drove the German fighters off. With hydraulics shot out and no brakes, he slowed the aircraft after landing by deploying parachutes from the waist and tail gunner's position. Six of thirty-nine aircraft on the mission were lost, two because of a mid-air collision.

Harry Carroll upper left and B-17 crew before a mission

I often wonder what Harry said to his men before such missions. How did he prepare them for such mayhem? On my desk at home, I have a picture of him with his B-17 crew. Everyone else is staring straight at the camera, some are even smiling, but Harry is looking off in the distance, as if he and he, alone, was contemplating the words written by General Bernie Lay in the book *12 O'clock High.* "Fear is normal . . . stop making plans . . . consider yourselves dead . . . then it won't be so tough."

This is physical courage defined, feeling fear yet choosing to act, an example of what I would later refer to as "Tough and Competent."

Harry was all of this and remarkably, a poet. "The other day while flying in my 'Fort' away up high, I saw some cloud formation . . . shaped like 'Islands in the Sky.'"

Like many who returned from that war, Harry's immediate postwar life felt strangely incongruous. At first, he tried to get a job as a commercial pilot but the airlines told him that they had all the pilots they needed. He eventually landed a job at the Mars Candy factory on Oak Park Avenue, just a short distance from his Chicago home. Built on sixteen acres, it was purportedly the largest candy factory in the world. Just months from the heroic air war that defeated Nazi Germany, Harry worked the night shift watching Milky Ways roll past him on a conveyor belt. As you might imagine, that did not last long. Utilizing the benefits of the GI Bill, Harry studied aeronautical engineering, first at Cal Tech and then the University of Washington where he graduated in 1951.

Life seemed like a hobby to Harry; he was always in search of ways to grow, learn, and teach. As an Eagle Scout leader, he led the "grand portage" canoe trek through the twists and turns between lakes along the Minnesota and Ontario border. He acted in dinner theatre and volunteered for the Outward Bound winter mountain wilderness survival where participants develop trust and organizations learn to adapt, grow, and succeed. He said, "I always volunteered, in that way I was the first with the opportunity to do something so that I did not have anyone to blame for how things turned out."

Harry would go on to work for Martin Marietta Space company, contributing his talents to projects like the Space Shuttle, the MX missile, and the Viking Mars Lander. But when I met him at McDonnell, Harry was the manager of a team of about twenty engineers, all of us assigned to reduce the data from test missions of various aircraft.

Harry never talked much about himself. He always preferred to talk of mountains to climb, challenges to meet, and new worlds to explore. A sincerely dedicated friend, his belief in honesty, humility, compassion, empathy, integrity, and learning were fundamental to my own personal growth. In other words, Harry exhibited the consummate skills of a great leader. He possessed a silent courage, concern for his team, and an unbridled enthusiasm for every endeavor he undertook. He was the most inspirational of all of my bosses who embodied the words of George Bernard Shaw, "I want to be thoroughly used up when I die, the harder I work the more I live. . . . Life is no 'brief candle' to me. It is a sort of splendid torch which I have got hold of for the moment; and

I want to make it burn as brightly as possible before handing it on to future generations."[2]

Quite simply, Harry was everything I wanted to be in life.

3

FLIGHT

Great pilots are made not born. . . . A man may possess good eyesight, sensitive hands, and perfect coordination, but the end result is only fashioned by steady coaching, much practice, and experience.
—*Air Vice Marshal J. E. "Johnnie" Johnson, RAF*

1955

After preflight training at Lackland Air Force Base (AFB), near San Antonio, Texas, my flight training began at Spence Air Base in Moultrie, Georgia. Spence was one of nine bases that continued to use civilian instructors for primary training. Pilot training Class 56M consisted of 104 students divided into two squadrons, Polecat and Panther. Instructors were assigned four students.

My student compatriots were Second Lieutenants, Vetter and McCullough, and Aviation Cadet Fass from the Netherlands. Our primary flight instructor was Mister Coleman, Panther 25. His full name was Morris Jackson

Coleman but as a sign of respect, instructors were always addressed as "Mister." When we graduated, he and his wife Barbara had a steak and beer dinner celebration for us where he introduced himself to us as "Jack." It felt like a rite of passage and indeed some of the guys began calling him Jack, but not me. Even when I interviewed him for this book, I could only address him as "Mr. Coleman."

"Mister" Coleman, Primary Flight Instructor

Born in 1929, Mr. Coleman was a shade too young for World War II but served two stints in Korea before leaving the Air Force for a career in aviation instruction. When I knew him, he was as thin as a rail. He tended to walk slowly, in a kind of shuffle, perhaps because he had spent so much time encumbered by a seat pack parachute that kept hitting his thighs as he walked. Pilots wore parachutes to the aircraft while students hung them on their shoulder. He was a chain-smoker and, at our first meeting, suggested that it was a good idea for us to also smoke Lucky Strikes. Mr. Coleman was a Southerner, through and through. Born and raised in Greenville, South Carolina, he could trace his Dixie ancestry on his father's side back to the 1600's. Listening to him, I discovered Southerners don't talk like people in Ohio, and are very proud of their complicated heritage. He claimed to have had a high school history teacher who taught that the South had won the Civil War.

Training consisted of a half day in the classroom and the other half on the flight line. My college background facilitated a rapid adaptation to the classroom, navigation, weather, flight planning, and aircraft systems. Flying also came naturally to me: takeoffs, stalls, spins, unusual attitudes, and aerobatics were a joy. However, approaching the ten-hour solo Go/No Go, the landings were giving me a problem. The military approach to landing used a forward slip, by establishing an aiming point on the runway as a target, bank into the wind. When the target starts moving apply opposite rudder to keep aircraft nose oriented to the target. Then use the elevator and throttle to control descent rate. For four flights, Coleman and I flew landings. Students always knew when I was practicing landings in the pattern. I would control the rudder and aileron, making the crosswind correction alignment, and Coleman would control throttle and elevator for the descent to the runway.

Separating my landing problem into two parts, crosswind control and descent solved my problem. As I became proficient in crosswind control, I could feel Coleman easing his hand off the stick and throttle during the approach. On the fourth day, after a few landings and 13:08 flight hours, Coleman climbed out of the rear cockpit. It was time to solo.

Mr. Coleman was more than your average instructor. Having taken coursework at Florida State University on "The Psychology of Instruction," he used techniques learned there to augment his natural skills. That day, alone in the airplane, I experienced the freedom of flight in its purest form. In the hours preceding my solo, Coleman had crammed a lifetime of his experience into me. He provided the motivation, developed the skills, and then forged my confidence to leave the earth behind.

Every day, he built his student's abilities to concurrently perform many tasks while controlling the aircraft, flying the prebriefed plan, and observing weather while always maintaining airspace situational awareness. Coleman was not just a teacher, but *a mentor*: developing skills, discussing his time in the service, and defining the relationship between pilots and the ground crews. He also set the standard that in life "There should never be such a thing as 'good enough.'" He was the leader I emulated in executing the flight program. High up in the sky over Georgia, he prepared his students to do things unnatural to a human; that is, to climb into an aircraft, enter a stall, and kick it into a spin knowing that if you did not execute the proper recovery, you would have to parachute from the plane or you would die. His instruction provided the confidence and skills to face fear, challenge it, and then ultimately push it aside.

Those who do not know fear are missing something. Controlled fear is useful. It sharpens the senses, triggers an adrenaline rush, and compels action.

Several times in my later career as a Flight Test Engineer and at NASA, I was assigned responsibilities for which I was not ready. I thought of Coleman's words when it was time to solo: "If the student says I'm not ready when I say he is ready," then I say, "That is all I need to know, don't come back to me. If you can't go when I say you are ready to go, then you can't go when you tell me you are ready." Back then, I had difficulty understanding what he meant, but later in life I worked with men and women who on occasion refused to accept new responsibility, and it was then that I understood what Mr. Coleman was trying to tell me. It is natural for a person to desire staying in their comfort zone, but a strong leader often knows the person better than they know themselves.

Forty hours into the program, the roar of the Cyclone 800 engine on the T-28 "Trojan" aircraft signaled the onset of more aggressive and demanding training. In the 1960s, the Trojan was used as a warplane carrying guns, rockets, and bombs for counter-insurgency missions in South East Asia.

With one hundred flight hours, aerobatics, instruments, and daytime cross-country flights completed, I moved into night solo training. On September 19, 1955, I began a 360-mile solo night cross-country over the swamplands of Georgia. At the first checkpoint, I made the radio call and, looking down, changed radio settings while beginning the turn to the new heading. Looking up, I saw the city lights of Alma, Georgia, eighty miles northeast of Moultrie, appearing like stars, filling the windscreen in all directions. For a fraction of a second, I stared in disbelief then realized that I had rolled upside down and was now diving in a "Split S," a maneuver you see fighter pilots assume when trying to escape a dogfight. With engine noise increasing, I throttled back and rolled to what I believed was an upright position, but now there was nothing but blackness in all directions: no stars, no lights from Alma, nothing, plus my airspeed increasing and altimeter unwinding. I pulled back and felt the aircraft shudder approaching a stall. I had induced vertigo when I looked down during a turn to change my radio setting, triggering the sensory system of the inner ear to send confusing signals to my brain. I finally did the right thing: buried my head in the cockpit and recovered by instruments, returned to level flight at a couple thousand feet, and started a climb back to altitude for the final two checkpoints. The vertigo never abated. Upon landing, I emerged from the cockpit completely shaken. Mr. Coleman's basic instrument training had saved my life.

The next day, as darkness approached, I realized I must conquer my fear or wash out of flight training. Chain-smoking, I delayed going to the aircraft until I heard them testing the loudspeakers for Saturday's parade by playing John Philip Sousa's march, "The Stars and Stripes Forever." The music stirred me to my purpose and parachute in hand, I strode to the plane and aced my second night solo. From that day forward when I faced a challenge, "The Stars and Stripes Forever" played inside my head and I pushed forward. My final flight with Mr. Coleman was a "buzz job," flying low and fast along the pines and backroads of Georgia. We met many times to reminisce in the coming years.

***

As quickly as I had arrived at Spence, I was soon off and began my jet training at Laughlin Air Force Base in Del Rio, Texas. My jet instructor was Lt. Wendell Dobbs. Less than a year earlier, he too had been a Laughlin student, he was only two years older than me. Although we were all equal in grade, he was our "Fearless Leader" and in more ways than one, the 56M Pilot Class was divided into an A Flight and an F Flight. The instructors in F Flight used a unique "F" call sign for their flights: "Flapper," "Flicker," and my flight had the best one—we were "Fearless."

Lt. Wendell "Quack" Dobbs,
"Quack, if you are airborne . . . let's hassle!"

Dobbs was a farm boy from Jacksonville, Illinois, who joined the Air Force right out of high school and was sent off to the Korean War. At Kimpo AFB, just outside Seoul, he undertook "Lamplighter" missions, dropping flares from airplanes to alert the frontline on the location of the enemy. He also worked as a mechanic servicing Douglas C-47 "Dakota" aircraft. After the war, he returned to Illinois where he earned a degree in aeronautical engineering. By the time we met, he was on his second Air Force stint, but Dobbs' best years were ahead of him. First, he worked as a test pilot for Aero

Commander, an aircraft manufacturer in Norman, Oklahoma, and then for thirty-five-years at American Airlines where he developed Extended Range Twin Operations (ETOPS). The rules and procedures for extended operations permitted twin engine jets to *operate on long oceanic routes* that had been previously reserved for tri and quad engine jet aircraft. His work earned him the title "Professor" of "The University of American Airlines," and in 1989, he was given the honor of opening Alliance Airport in his adopted hometown of Fort Worth, Texas. He helmed a Boeing 757 where he shared the cockpit with H. Ross Perot, an Alliance co-investor and future presidential candidate.

It took a bit of time for me to get used to Dobbs' approach. Whereas Mr. Coleman had followed a steady and deliberate instruction path, Dobbs was something of a loose cannon known by his nickname of "Quack" (I never figured out the reason behind this moniker). My initial impression was he did not possess Mr. Coleman's talent, but he made up for it in pushing the performance boundaries of the aircraft.

Dobbs had three students in Fearless Flight: "Sully" (Jack Sullivan) and "Stash" (Stan Kurzeja), and me. His instructions were rudimentary and never repeated. He was short-tempered but when he pressed you and you responded to his pressure, it showed how gifted he was as an instructor. His rating system was simple, *two* strikes and you were out.

There was no syllabus with Dobbs so the Lockheed T-33 cockpit was a busier place. Like many of the jet instructors at Laughlin, Dobbs was a fighter pilot wannabe. As such, we did less "by the book" instruction and more what he wanted to do which, as it turned out, was a lot of combat-related training. His teaching approach was to be less your friend than your adversary, always pushing and always demanding persistent maximum performance. With the jet solo came the ceremonial white scarf and, over the next few weeks, scarves became prominent throughout the squadron. Almost daily, at the end of a training session with Dobbs in the cockpit behind me and about twenty minutes of fuel remaining before "Bingo"—the fuel minimum call—I would hear from the radio, "Quack . . . if you are airborne let's hassle." Dobbs would reply "Roger," set a rendezvous point and shortly, a fight would begin. Aircraft flight manuals provide the safe limits for all flight conditions that are accompanied by written warnings and cautions. In those daily sessions with Dobbs, I learned how to fly the Lockheed T-33 jet trainer on the edge of the published aircraft safe limits. I was learning by doing. I could feel Dobbs's hand on the stick, when approaching a flight boundary.

Just a few days before the mid-program cross-country, Dobbs really chewed me out after a flight and he did so in front of Sully and Stash. I was embarrassed and responded angrily, "Dammit, I am doing my best!"

That clearly was not what he wanted to hear. He looked squarely at me and his voice, taking on the guttural tone of an angry dog, said, "You only *think* you are doing your best," and he was not done. The next day, during the cross-country flight Dobbs picked up where he had left off. "Lieutenant," he said to me, "why did you get into flight training?"

I again responded with what was the wrong answer. "To fly and get my wings."

He spit back a challenge. "Is that enough?" There was a brief period of silence that seemed to last for an hour, and I could feel nervous perspiration beading on my neck. "Your wings are not enough," he said. "Set your goal to be the best." He explained that every pilot training at Laughlin was hoping to land one of two Fighter Weapons School slots available at Nellis Air Force Base in Nevada.

Today, Nellis is "Top Gun," home base for Red Flag, the contested combat-training exercises of the United States and our allies on the Nevada test and training range. Dobbs continued, "There, you will be competing with the best in the Air Force. You have the skills, and I am going to challenge you to use them." There was another pause, this one more welcome, as he was paying me a compliment. "From now on, your goal is to get one of those slots, but to get to Nellis, you are going to have to beat the top student in 'A' Flight." Dobbs and another instructor so wanted us to succeed that they took a page out of the Nellis training program and included it in our jet training. After we completed the required flight fundamentals, we focused on formation and air combat maneuvering (ACM) on almost every flight.

Often when I was scheduled to fly with Dobbs, I was assigned to another instructor while he would fly on my wing in a different aircraft with a student. The planned flights became even more demanding with the hassles finishing more of the day's scheduled training. Soon, I had my first inflight emergency: major engine instability followed by a flameout. I made a restart followed by another power loss, but I made the right calls and nursed the aircraft back, flying a flame-out approach at Laughlin.

Formation flying separated the pilots into two groups: those who flew good formation and those who didn't. Only the best formation pilots had a shot at fighters, and I was flying good formation. Dobbs was an incredible coach. He appreciated the physical "touch" demands of formation and the guts to get in so close that the formation became as one. After landing, the sheer physical exertion of formation showed in the sweat rings in our flying suits, the matted hair, and the indent of the oxygen mask on our faces. "Fights On" were the words I now lived for each day, but daily, I would come home exhausted. *If this was just basic training, what would advanced training be like?* In the end, I appreciated what Dobbs had done for me. I graduated from jet training as a "Distinguished Pilot Graduate," first in flying, third in academics.

To my surprise, I found that the pilot I was competing against on A Flight for the Nellis assignment was my best friend and a former Air Force Flying Sergeant Lt. Anthony Zielinski. As it turned out, he graduated second in flying and first in academics; we got the two slots for the Fighter Weapons School at Nellis.

My commitment to be the best paid off and set my standard for the rest of my life. This was the first time in my life someone recognized I "had more to give," and that I needed to set my goals higher.

I still wonder how Dobbs arrived at this recognition and what he saw in my performance, demeanor, words, or attitude that said, "This guy needs to set his goal to be the best." Now I know that being the "best" meant being both accomplished—that was the "Competent" part—and being "Tough" enough to take on new challenges.

On April 27, 1956, Zeke and I received our pilot wings and the two assignments to the Fighter Weapons School at Nellis. There we would fly the world's best and fastest fighter aircraft in the world. How could it get any better? As it turned out, Dobbs prepared me well for the transition to Nellis and the Fighter Weapons School. His aggressive training approach and his frequent air combat sessions provided me with the confidence and "feel" to operate in the boundary region of my aircraft's performance. Equally important, however, was developing the ability to observe and anticipate my opponent and remain on the offensive.

In my later life at NASA, I would often issue a challenge to those I believed had "more to give" and even today when speaking to youth groups, I tell them to "Dream, Aim High, and Never Surrender. Then, as you get closer to your goal, aim *even higher.* Just keep moving the goal forward so you are never satisfied with your present skillset."

Back in 2000, I was at an air show at Daniel Webster College in New Hampshire signing copies of my first book, *Failure is Not an Option.* At the next table was Gabby Gabreski, the World War II and Korean War ace. As we began to chat, he said, "Audacity is the hallmark of a fighter pilot. You must be aggressive, or all the other skills are wasted." I remembered I had learned that skill from Dobbs and he had made me ready to aim even higher!

4

THE CREED

I am an American Airman. I am a Warrior. I have answered my nation's call.

—The Airman's Creed

The Nellis Air Force Base Fighter Weapons School (FWS) was where the flame of fighter aviation was kept alive in the mid-50s. It was a metaphoric Valhalla for the top two percent who graduated from jet training, the mythic Norse palace roofed with the shields of noble warriors. In his book, *Boyd*, Robert Coran describes the FWS as "More than the top finishing school in the Air Force, it is the Temple of fighter aviation . . . its graduates are priests, the instructors high priests."[3] Our textbook was a newsletter written one year earlier by a Korean double Ace, Major "Boots" Blesse, "No Guts, No Glory." Its timeless message could have been written in 1917 for American ace Eddie Rickenbacker's "Hat in the Ring" Squadron and for those who flew the Hellcats, Mustangs, and Thunderbolts of World War II.

The F-86H Sabrejet was an improved version of the F86 that achieved a MiG kill ratio between 6:1 and 10:1

over Korea. (Kill ratio is a common yardstick of fighter performance which indicates the number of enemy aircraft destroyed for each U.S. fighter lost.) The F-86H entered service in 1954 as a fighter-bomber with a more powerful engine that produced 8,900 pounds of thrust which provided increased climb rate, airspeed, and payload and with greater flight range. Zeke and I were assigned to the F-86H/F-100 (Super Sabre) training program. Because of competition for the F-100 with instructors, the Thunderbird team, and design issues that triggered a violent pitch-up and stall called the "Sabre dance," the F-100 was eliminated from the 56M class curriculum. With no simulators or two-seat trainers, the first flight was truly "solo." Training consisted of blindfold cockpit time, extensive classroom training on the aircraft, performance, handling qualities, emergency procedures, and one of "time compression"—the events that occur in seconds in the "first flights."

At the FWS, we were not only learning to fly the most advanced aircraft in the world, but how to use it as a weapon. Flying with armed guns and external ordinance underscored the change in mission. At the completion of every flight, we debriefed reviewing available gun camera film, range reports on strike effectiveness, and critiques from the pilots involved. After three weeks of classroom instruction and cockpit drills, my tenacity paid off. On July 5, 1956, I was scheduled for two successive flights. For over a decade, I had carried this dream, sustained it, studied it, fought for it and now I was ready to begin the life of a fighter pilot. The first flight was pure exhilaration, seemingly only seconds long, it passed in a flash. Takeoff, gear and flaps up, the instructor called out a heading, and it seemed we were at 30,000 feet in an instant. I now understood the expression, "Time compression . . . and being behind the aircraft." My thought process had to reset from my early jet training to fighters. All too soon it was time to come home. Entering the landing pattern, the instructor called out "Airspeed . . . speed brakes, flaps, and landing gear." Moments later, my first landing was a grease job. An indescribable joy ended the second most wonderful day of my life.

The second flight involved higher airspeeds with rolling and turning maneuvers, feeling the pressure of the G-suit at 4 to 5 G's, (five times earth's gravity) and dives to get the "wing heavy" feel of the control surfaces at .95 Mach (95 percent the speed of sound.) The second landing went well with only limited coaching. Then it was back to the classrooms.

Daily flights brought proficiency during high-angle bombing, skip bombing, strafing and rockets, plus day and night formations. September was spent in the classroom preparing for the next training phase straight out of Blesse's book on combat formations. We reviewed combat film from Korea, briefings on formations, teamwork in offensive and defensive formations, rapid change of leadership roles, and communications. The lessons on the role of a wingman in providing another set of "eyes" in protecting the leader when attacking or being attacked became a key function in my subsequent leadership approach. Dobbs had taken a page out of the Nellis manual and prepared me well for this critical phase of training.

***

Following graduation from Nellis, I was assigned to the 354th Fighter Day Wing, the top fighter group of World War II. In seventeen months of combat, the group had flown 18,334 sorties, shot down 701 aircraft, and destroyed another 255 on the ground. The 354th Base Commander at Myrtle Beach was Colonel Francis Gabreski, a double Ace from World War II and Korea. When I reported for duty in December 1956, there were no Sabres or F100s on the flight line, only a couple of T-33s and six F-80Cs which were relics of the Korean war. The F-80 Shooting Star was our nation's first mass-produced jet aircraft capable of 500 mph in level flight. The cockpit was so small, you did not climb in; you just put it on. The airspeed indicator was in MPH and, except for instrument flying, was a joy to fly.

On December 22, 1956, I drove to meet my mother in Toledo and introduce her to Marta Cadena, the young lady I had been dating while in jet training, who had pinned my wings at graduation. Marta's trip was by air from Eagle Pass, Texas, and subsequently by rail. Flights grounded by snowstorms made our planned engagement on Christmas Eve a challenge. Walking back to the Mercy Hospital nursing home in the snow after mass and our engagement was a beautiful way to start our life together.

My future wife, Marta Cadena

In January 1957, I completed F-100 ground school at Langley AFB in Virginia. In February, the 354th Group commander Colonel Hackler initiated F-100 ferry operations from California. To requalify for transition to the F-100, the 355th deployed to Seymour Johnson AFB, North Carolina, on April 8 for a recheck on the F-86H. Two weeks after completing recheck and returning to Myrtle Beach, I received orders assigning me to the 58th Fighter Bomber Wing (FBW) at Osan, Korea, flying the F-86F Sabre. My new assignment quickly got around the squadron and the following day, Captain Love, my flight commander, said, "We know you got orders to Korea, but the boss said to check you out . . . tomorrow's the day."

In the 1950's, there were no simulators or trainer versions of the aircraft. The F-100 was a "Bear," immense and the first fighter with supersonic capability in level flight. On takeoff, you were airborne and over 160 knots by the runway end. The aircraft would become the principal ground support aircraft in South Vietnam, flying over 360,000 sorties in eight years. In his book, *Songs from a Distant Cockpit*, "Hun" pilot John Schultz wrote a poem about the F-100 which he titled "Supersabre: Our First-Class Lecture."[4] It begins:

Supersabre sits in the velvet night
Awaiting day, armed for fight.
"The men who designed it planned from the start
To put fire in its mouth and steel in its heart"

My first F-100 flight was June 15, with almost daily flights through early July, finishing with five tactics (air combat) missions. I will never forget 16,000 pounds of thrust as the afterburner kicked in and the incredible power at my command. The cockpit was vast, visibility superb, and even on the first flight, it felt solid but I knew it demanded respect.

The good, solid feel of the drag chute deploying on rollout augmented braking on my first landing.

Closing in on reassignment, my 15th flight was scheduled with two other aircraft. Once airborne, an additional aircraft asked for permission to join formation. I soon heard, "Take my wing . . . I am lead. Follow me," and I started flying as a wingman on the unknown aircraft. Our aircraft were so new that they did not have pilot's names or squadron markings; however, I did have a clue about the pilot of the visiting aircraft. During debriefing, Captain Dillon signed off on my proficiency check and cleared me for operational training. He commented that I had been flying on Gabreski's wing, America's top living ace after he joined formation. Mine was a brief fling with the Hun, but it never blossomed into a full-fledged love affair. There is an Alfred Lord Tennyson quote "'Tis better to have loved and lost, than never to have loved at all." I hoped that future assignments would again return me to the cockpit of a "Hun."

I had just finished my high school junior year when the Korean War began in June 1950. My only dream was to fly, becoming an ace high in the sky over Korea in a Sabre. Now, years later, we were mired in a Cold War, and I was there to fight if needed. I was assigned to the 58th Fighter Bomber Wing (FBW), at K-55 (Korean War airfield designation of Osan-Ni airbase).

"The Fighting 69th" Fighter Bomber Squadron, K-55 Korea
1957

The 58th had three squadrons: 310th, 311th, and 69th. I was assigned to D Flight commanded by Captain Forsyth, the 69th flight commander. The 58th Wing rotated squadrons to Taiwan on a routine basis, supporting the 1955 Sino-American Mutual Defense Treaty. Two weeks after arriving in Korea, we were deployed to Tainan AB, Taiwan, home to several Nationalist Chinese F-84 Squadrons. We arrived August 31 and settled into the bamboo hooches (shacks). Typhoon Carmen, with winds up to 125 mph, grounded us for a week after arrival. We spent two of the days sitting two pilots abreast on the top bunks in our hooch in drenching rains, with occasional visits by frightened monkeys and snakes searching for cover.

After exercising authentication procedures, we began four-ship patrols over the Taiwan Strait. Air combat tactics and practice intercepts with Republic of China (ROC) controllers were part of the daily drill. The Rules of Engagement (ROE) over the Strait had the ROC as the first to engage any Chinese communist aircraft if they showed up. The Taiwan deployment, frequent flying, daily patrols, and joint operations with ROC controllers fostered the close-knit team skills and confidence of a flying, fighting unit.

At the end of our deployment, I was advised I would be assigned as a flight lead on our Korea return. My test as flight lead and my weather-flying abilities came earlier than I expected. After attaining cruise altitude *en route* to Okinawa, Japan, Captain Carroll rocked his wings and with hand signals indicated I was to take the lead. Carroll, without communications, moved to my left wing, and I had two replacement pilots as an element on my right wing. The pre-takeoff weather briefing indicated we would hit severe weather with thunderstorms and heavy, blowing rain during descent at Okinawa. With sunset rapidly approaching, I could see lightning illuminating the clouds ahead with tops appearing at an estimated 30,000 feet. I briefed the flight on Kadena frequencies, ground-controlled approach (GCA), missed approach, and the procedure if we got separated.

I turned the cockpit lights up high to prevent blinding from the lightning and initiated descent on the outbound heading. Passing over the Kadena beacon at 20,000 feet, I continued the descent, starting the turn to the inbound heading at 10,000 feet. Communication silence indicated the flight was with me. Midway through the turn, I heard the welcome voice of the GCA radar operator advise, "Slightly high . . . continue to zero-five-two . . . rain showers, left crosswind." After completing the turn to the runway heading, the confident voice of the GCA operator continued calling heading and glide path corrections. After breaking out between 500-700 feet, I called flight, "Separation . . . landing long." This provided runway space for following aircraft. The "Follow Me" truck met the formation at the runway end in a steady downpour. After engine shutdown and canopy opening, an airman quickly held an umbrella over my head, not doing much good. After what seemed a long time, but must have been only a few seconds, he said, "Lieutenant, aren't you gonna get out?" I was just glad to get my flight down alive.

***

The Korean peninsula is divided by the Demilitarized Zone, a line that runs diagonally across the 38th parallel (38 degrees north of the Earth's equatorial plane) about twenty-five miles north of Seoul, South Korea, on the Western edge. There are two critical boundary lines: the Demilitarized Zone (DMZ) and the Joint Operations Command (JOC) line five miles south of the DMZ. While just lines on the map, these are the critical boundaries. Aircraft overflight would invite North Korean anti-aircraft fire.

"Declare" Flight Lts O'Neill, Ball, Shover, Flight Lead Kranz

There were thirty-four pilots in the 69th Fighter Bomber Squadron (FBS) commanded by Major William Nacy. Captains were assigned to Flights A through D, and I was assigned to D Flight, commanded by Captain Tom Forsythe. Nine pilots, mostly Second Lieutenants fresh from flying F-84Fs at Luke AFB, Arizona, were sent to fill out our ranks. Returning from Formosa, Taiwan, I was assigned three Second Lieutenants: Ball, Shover, and O'Neill for my new flight.

After briefings on squadron policies, Korean airspace, and strip alert, I began formation and tactics evaluation with my flight. An element of two aircraft, lead and wingman, is the basic fighting team. Two elements comprise the effective fighting unit. Training develops the instinctive ability to transition between offensive maneuverability and defensive mutual support and is aggressive and unforgiving. Following the words "Fight's On," the flight separates, turns outbound, and returns in a heads-on pass closing at about 800 mph. Most engagements will last for an intense 1 to 3 minutes or occasionally longer, until an altitude or low fuel call.

***

Second Lieutenant Dale Ball, call sign "Bingo," flew as my wingman for much of the time in Korea. Ball, born in 1934, was raised in Michigamme, Michigan, population 247 on the Northwestern end of Lake Michigan. His parents, Norma and Morris Ball, owned the Mount Shasta Lodge which was used for some scenes of the movie *Anatomy of a Murder*. Graduating as a Mechanical Engineer from nearby Michigan Technical University in 1955, he was subsequently commissioned in the Air Force. He received his wings at Laredo AFB, Texas, in 1957 and was assigned to the 58th Fighter Bomber Wing, K-55 Korea. Ball was shorter than the average pilot at about five feet, seven inches and 155 pounds, cocky, and had a perpetual grin. He was a hunter and fisherman and liked to tell stories.

Ball and I communicated until New Year's Day 2021 when he died because of COVID-19. He was a writer and often during our talks, he would read from letters he wrote to himself late at night while caring for his sick wife. They had interesting titles: "Combat Ready," "Cannon Fodder," "River Ride," and many others. I occasionally reread his letters since I am at the time in life where I no longer have someone with which to share experiences.

***

The wingman's duty is to look around and provide eyes to cover the sky, "checking six" in the lead aircraft's rear. A wingman must never lose his leader and must maintain a "fighting" position, sliding inside, outside, and in trail with the lead aircraft regardless of the lead aircraft's maneuvers. Wingmen must possess the same level of aggression as the leader and must be close enough to the lead to avoid becoming a target for an attacker. The three-dimensional aspects of the airspace, offensive and defensive tactics, and rapid changes are extremely demanding of piloting skills and of energy management, trading airspeed, and altitude working to retain the offensive. Aircraft performance varies so leaders must provide wingmen a slight throttle margin so they can keep up. Gabreski was often called too aggressive by wingmen, while others considered him a mentor, "My kind of fighter pilot." Ball was "My kind of a wingman." Flight leads check out potential wingmen during aggressive tail chases, trying every possible maneuver to shake them loose. I considered Ball unshakable.

In today's complex, fast-moving, and challenging world, I believe that top-level leaders should seek out and utilize individuals within their organizations with the character to assess and provide a qualitative assessment check on the performance, effectiveness, and integrity of their leader. I have utilized a "wingman" in every aspect of my professional life, often calling them my conscience.

The second leadership lesson learned in flight is one of energy. To a fighter pilot, dynamic energy management, both kinetic and potential, in air combat maneuvering is the principal asset in his hands. Altitude can be converted for airspeed and the reverse is also true. A pilot is constantly trading one for another.

The success of virtually every organization depends on leadership skilled in effective "value judgments," such as timely tradeoffs in time, personnel, financial assets, functional assignments, and programs and projects.

My Korean tour was not without losses. Our flight lost a pilot in gunnery because of a wing structure failure when pulling off the target. His bunk was across from mine, and I performed the sad duty of packing his gear for shipment home. Additional squadron losses came from a drop tank that separated on pitchout for landing and another on takeoff. Our aircraft were leftovers from the Korean War and were "officially" limited to 3G. We did not observe the limit and punched out the G reading on the accelerometer at engine shutdown after every flight, saving our crew chiefs overstress inspection time.

I was assigned as a Forward Air Controller (FAC) to the 7th Infantry Division (ID) in February 1958, and assigned to Major Witt Barker, G-3 Air. The unit was stationed as a blocking force at Camp Casey north of Seoul. FAC training was "learn by doing," as there was no FAC manual or even a good set of notes. I was provided an artillery compass and old, worn maps to plan and direct air strikes. The good news was the Army provided good cold weather gear; the bad news was the cold. I was flown by helicopter or Army light aircraft to reconnoiter the thirteen FAC locations ten to fifteen miles south of and spanning the Demilitarized Zone.

Operating from a ¾-ton truck with radio gear, the Army team consisted of a driver, radioman, and wireman. Wire was strung from spools on the truck to my selected hilltop observation location. The drivers were exceptional, navigating along narrow, bulldozed icy mountain roads with minefields on the roadsides, usually in the early morning with blackout lights that used a diffused horizontal light beam. Air strikes were conducted against old North Korean bunkers constructed and installed during the truce period. The terrain was mountainous, and the flying was unlike anything we had ever experienced. My time with the 7th ID was one of the high points of the tour.

I had never played on a sports team so I was unprepared for what turned out as the capstone of my military flying career. The 69th Fighter Squadron, the last Sabre Squadron in Korea, was inactivated in June 1958. Prior to transferring our Sabres to the Nationalist Chinese, we flew a final mass formation, a "Diamond of Diamonds." I was one of sixteen pilots who united in an effort to say farewell to the Sabre and to each other. That day aloft against a deep blue sky, the squadron flew as one.

Team chemistry as I had never known before created a solidarity. We felt the emotion that in our time, we belonged to each other and to something great. It was that bond, that feeling that we were all involved with something together, that would last for me. Indeed, I later called upon these memories when building teams at NASA, the ones that were committed, together, to achieving "IT," the common goal.

On June 2, 1958, at Tainan Air Base, Taiwan, I secured the cockpit and closed the canopy of "My Darling Marta," the F-86F Serial number 524872 that had served me well. I knew my crew chief, Tom Gerdes, well and loved him like a brother. By tradition Gerdes named the right side, "Miss Judy." Our names were on the canopy rim. Thinking of that moment now, I am reminded of a line from Ernest Hemingway, "A man has only one virginity to lose in fighters, and if it is a lovely plane, he loses it to, there his heart will forever be."[5]

The nose art on the F86 Sabre dedicated to my wife

My time in Korea was unlike any other time in my life. It was deadly, harsh, lonely, and yet, a great time that provided a new dimension to life. I observed, worked, and made close friends with individuals of different backgrounds and upbringings. We became uncommonly close as we lived in four uninsulated corrugated sheet-metal hooches, enduring the discomforts faced on a daily basis. A short walk brought us to the primitive latrine and, if we were very lucky, the Army brought in a trailer every couple of weeks for a hot shower.

In the words of Bingo, my wingman, our hooch was cold, the operations building was cold, the planes were cold, and the latrines were cold. Christmas was in a cold hangar, the USO Troop was cold, but Bob Hope, Jayne Mansfield, and the Les Brown Orchestra were outfitted in parkas. There was no rank as we lived together, shared together, celebrated together, and complained together.

We were given a mission as a Cold War outpost, five minutes flying time from North Korea. On Christmas Day 1957, after singing "Silent Night," we returned to the hooch, celebrated with Scotch whiskey or Asahi beer. After rereading recent letters, the memories of home and the friend we recently lost on a strafing run receded into our alcoholic memories. Johnny Cash, Elvis, Buddy Holly, Sinatra, and Patsy Cline on Armed Forces Radio soon competed with the snores of exhaustion, ending another day. A final call, "Shut the damned thing off!" brought silence to the hooch. The most difficult time in life comes when there are few remaining with whom to share these poignant memories and emotions.

The closeness within the squadron taught lessons in leadership, adaptation, and comradeship. I learned of sacrifice, leaving home, family, familiar surroundings, and ways of life, and I came to appreciate what I previously took for granted. Visiting Korean orphanages, passing villages with gaily-garbed children, with only women and a single Papa San provided a clear visualization of the price of war on society that would take a generation to recover.

> Leadership is ultimately about responsibility. History shows how the demands of leadership changed from one era to another. Many of those I flew with, our squadron and group commanders, were trained in the past and had a different perspective on the present than me. They were aware that leadership is partly luck and chance, that simple little things provide the difference between success and failure, life and death. The lessons of their example of leadership, even their selection of words, brought forth the values they held close, and I learned from them.

In Korea, I was surrounded by men, good solid men, that I trusted with my life. Squadron commander Major Nacy and Captain Tom Forsythe; my flight commander; and my pilots, O'Neill, Ball, Shover, were men of honor and integrity. They were not only committed to a mission but to their team. Leadership, in many ways, is derived from close observation of leaders; from witnessing the "Tough" and the "Competent" in action. Nacy and Forsythe were firm and resolute in their commitment to the mission, their flight leads were trained to execute, the airmen possessed the qualities to serve, and they provided the trust that kept us flying. Above all, our leaders exuded the optimism that their squadron would not fail.

Alert crews met us at the aircraft at 3:00 a.m. on frozen mornings and maintained them long after sundown. When we got the klaxon signal horn to scramble and I taxied out, my crew chief, Gerdes's hand salute signaled all was well with my aircraft. I had absolute trust in the team that packed the parachutes, maintained my oxygen systems, serviced the aircraft, briefed the weather, armed the weapons, provided the transportation, and even the mess crew who made sure we had hot coffee with our ham, eggs, and grits. Teamwork got us airborne and got us home.

The Fighter Pilot's songs carried a meaning and the words, while coarse, spoke of life, death, beautiful women, and of a relationship that only men and a few women will ever fully understand. Words of the Airman's Creed come to mind today:

I am an American Airman. I am a Warrior. I have answered my Nation's call. . . .
I am an American Airman. Wingman, Leader, Warrior. I will never leave an Airman behind, I will never falter, And I will not fail.[6]

I remembered Gabreski talking of his commands:

One of the biggest jobs was to keep the men motivated under adverse conditions. I had a special, red-painted

Jeep and, when I drove it, the men would know I was out there with them every day, even if I did not stop by each revetment every day.

I now knew what Gabreski meant. I learned that the teamwork component of leadership is learning about the world through a different set of eyes. Now more experienced in many ways, Korea, Japan, and Taiwan provided my new vision. In many ways, when I left the U.S., I was leaving one life and entering a new one. Preparing for return to the states, I felt shaken by the challenge of returning to the life I had left behind.

There was an intense feeling of loss, and I wondered if I would ever again feel the love and passion for my work and the team chemistry of a fighting organization as I had with the 69th and the Sabre?

I knew I would have to reach deep down inside to find the resilience and the toughness that carried me before. Like Harry Carroll, I had traveled the United States, visited foreign countries, and acquired many skills but most of all, I was capable in every place of thriving in every environment. I had acquired the ability to deal with all people under any and all circumstances. I was now a different person than when I arrived in Korea and joined the Squadron. Maybe it was experiencing a unique organizational chemistry? The interplay of psychological, social, and emotional abilities that ignites the collective talent of a group to achieve something intangible, *team chemistry*.

While I still had much to learn, I felt I was no longer "the kid." I had transitioned to a responsible adult in my own right and I looked forward to flight test and using my new skills. The military provided training involving challenges, risks, and the need to step forward and confront events.

Military service demanded a unique, personal, and profound courage. A reckless defiance of the odds combined with the acceptance of accountability for an event. Those words had entered my vocabulary: *accountability, trust, teamwork, courage, toughness, and competence.*

I do not know why, but I felt I had come of age and it was time to forge my mark. I was going home to a wife, one who I had only known for three months, and a daughter, Carmen. A full year had passed, and I prayed we would once again feel the frenetic joy and passion of the springtime of our life.

5

HUMILITY

Knowledge will give you power, Respect will give you command.

—Bruce Lee

There were numerous teachers in my early career, some who moved in and out of my life before I even learned their names. For instance, there were many truck drivers I encountered while hitchhiking. They talked to stay awake and I was the guest in their cab, from Toledo to Cleveland, or Detroit for air races, and later from home to Parks College in Illinois. Every new journey provided opportunities to acquire a new perspective, expand knowledge, or just plain listen. I sometimes think that the many hours I spent in trucks provided the kinds of lessons I believe I would have received if my father had lived.

You could chart the nation's history through the stories the drivers told. Back then, they were one of two types: the old ones and the young ones.

The older men spoke of economic hardship. They were marked by the Great Depression and spoke of broken families, lost employment, and bouts of deep, abiding despair. I became an avid listener to stories about the pain of those times, stories that my mother had probably been too proud to share with me. Until I rode with them, I had never heard of the drought, the dust storms, and the black blizzards of the 1930s. These were the storms that choked humans and cattle alike, and turned the Texas and Oklahoma panhandle as well as adjacent states, into a dustbowl driving 2.5 million people out of the Great Plains states.

In contrast, the younger drivers had served in the war and talked about the many exotic-sounding places visited. Listening to them, you always sensed they edited themselves very carefully, that there were war secrets they weren't going to tell a young kid they picked up on the side of the road. It is a cliché about soldiers that they never talk about what really happened; that is, unless they are talking to someone else who had the same experiences. In later years, I developed the same mindset about some of my own experiences: you can only speak sincerely to those who have "been there."

During occasional long-haul trips back to Parks, the drivers would stop a short distance after crossing the Illinois state line and meet another truck. When they did, I would jump out and help them offload several heavy cartons to the other truck. At first, I suspected that these were cartons of cigarettes, and the maneuver was being done to avoid taxes, but the boxes were too heavy for that. After a couple trips of this sort, I finally asked a driver, "What's in the cartons?" and sporting a smug expression he told me of the "Oleo wars." It turned out that the cartons were filled with yellow margarine or "Oleo," which was outlawed in the dairy states. Oleo was originally made from beef tallow and is now made from vegetable oil. It had a kind of ugly, glutinous, white color in its natural form that looked a bit like grade school paste, so to make it a more appealing substitute for butter, the manufacturers would inject it with a yellow dye. Restaurants in particular liked to use Oleo because it was cheaper than butter. In response, the dairy farmers, eager to preserve the butter market, had convinced lawmakers in several states to outlaw "yellow" Oleo. So I was unwittingly a party to an Oleo "bootlegging" operation. It was an amusing lesson in interstate commerce.

I believe that by listening to the drivers, I developed a sense of empathy that improved my leadership skills later in life. In the Air Force, I served as counsel at a company punishment for an airman and while moving through leadership at NASA, I supported those with work, legal, financial, and emotional issues. If a collection's agent or a law enforcement officer showed up, I would not allow them to enter the employee's office area. Instead, I would personally escort the individual to the security office and have the encounter take place there. I would not let my employees be embarrassed in front of their units.

After leaving active Air Force service in August 1958, I returned to McDonnell Aircraft in St. Louis and worked as a production flight test engineer on the F-101A Voodoo. The Voodoo was a supersonic fighter aircraft designed as a long-range bomber escort. However, McDonnell was a different place than before and I became frustrated by my return to civilian life, especially by union demands that required "crew breaks" during testing and change of shift. I wanted to go for a flying position and noticed some openings in the F-101B Voodoo Auto Attack program at Holloman AFB near Alamogordo, New Mexico. The F-101B was a two-seat interceptor with a pilot up front and a radar operator, or "scope dope," in the rear. At that time, the Soviets were projected to fly at high altitude over the north pole. The plan was that the F-101B would accelerate to Mach 1, pitch up, and the radar system would automatically fire the missile at a predetermined range. The "scope dope," who was alternatively called GIB (for "guy in back"), acquired the target and set up vectors for attack. I looked forward to flying in that program.

Unfortunately, by the time I got to Holloman the Voodoo positions were filled and I was reassigned to the McDonnell XGAM-72 program.

The XGAM-72 Green Quail was a decoy missile launched from the Boeing B-47 and B-52 Strategic Air Command (SAC) bombers. It was equipped with folded wings so that four could be squeezed into the B-52 aft bomb bay. Then, after launch, the wings would open in flight. I suppose that is why it was called a "quail." To confuse Russian radar, the missile was capable of changing altitude, flight path, airspeed, and it provided an infrared signature that replicated that of the bombers.

The flight test program operated from a large, two-story hangar on the West field at Holloman, which was an oven in summer and freezer in the winter. The team consisted of about fifty mechanics and inspectors, twenty-three engineers, two secretaries, and it was led by a thirty-five-year-old Ohioan named Ralph Saylor. I felt alive again! The Holloman test environment was electric, packed with people doing new and exciting work daily. Every working hour, I was learning something new. The smell of JP-4 jet fuel, the whine of jet engines, and the supersonic booms that rattled the buildings brought me back to my element. The Army was testing surface-to-air missiles and while looking down range over the lava beds, I could see the rocket streaks coursing through the sky. An adjacent hangar housed the Fighter-Missile Test Branch and the latest jets from many friendly foreign countries.

Marta and I were thriving in marriage and soon, we were growing our family. The proximity to the mountains, the town of Cloudcroft, the Mescalero Apache Reservation, White Sands National Park, and the gypsum dunes covering the Chihuahua Desert were a marvel to behold. We were enjoying personal time, a second daughter, and a new phase of life.

When I arrived, Saylor was just forming the flight test organization. Daily for about two weeks, he would call a meeting in the hangar, introduce new personnel, and assign responsibilities. I knew little about Saylor then, but soon came to recognize him as a great leader with a unique ability to assess personnel, fit the right people to the right jobs, and stretch them to fill their assigned work. He was not a taskmaster, instead, he gave you the objective and guided you forward. Two qualities personified his relationship with his people: everyone who worked on the team was of equal importance and even when the team grew to be over one hundred people, he knew them all by name and knew the work they performed.

My time in college, at McDonnell, and with my mechanics in Korea, provided a solid knowledge of aircraft maintenance. After working the B-52 aircraft for a few days, the union pecking order for the contractors became apparent. The highly-skilled mechanics from Pacific Airmotive (PAC), the West Coast Union, were being used for the dirty work, to wash the aircraft, and to police the ramp while the Saint Louis personnel worked the missile hardware. I sent a proposal to Saylor recommending the assignment and the exclusive use of the PAC mechanics on the B-52. The proposal was readily accepted when Saylor and the shop foreman saw it as an end to the union bickering. While new to the job, I trusted my observations and experience to address an issue that was a waste of talent and time that needed to be solved before it impacted the test program.

Sometimes leadership is simply recognizing a problem and providing a clear and simple recommendation for a solution.

With the first Quail test launch pending, I had a lot to learn. Saylor's organization had four major functions: B-47 carrier aircraft maintenance (Boeing had the B-52 part of the job), Quail combined systems test/preflight, J-85 propulsion system test, as well as the instrumentation and central test procedures and equipment team. Each element had dedicated shop and inspection personnel and except for the J-85 engine test, the flight test engineers performed duties laterally across all functions.

Boeing B52D 56-695 flew in Vietnam and found a permanent "home" at the "Tinker AFB Heritage Museum."

At the beginning of my third week, I walked down the ramp with Saylor to the B-52. It was massive: 159 feet long, forty feet high, and with a wingspan of 185 feet. The fuselage skin was wrinkled like an old man's face, and the serial number 56-695 was stenciled on the fin. The Experimental B-52 Stratofortress had first been flown in April 1952 and the initial "D" model deliveries of 170 aircraft began in December 1956. Continuous jet engine and electronic upgrades would keep the B-52H version, originally manufactured in the Kennedy era, flying well into the next century.

After a formal introduction to the Boeing test team, Saylor and I returned to the aircraft. Near the open bomb bay, he offered his hand and simply said, "The B-52 and the PAC team is yours!" A handshake with Saylor was a contract sealing the assignment. He then turned and walked away. The Quail program was McDonnell's responsibility and the B-52 aircraft was Boeing's responsibility. The Quail used the General Electric GE J-85 jet engine. So, my PAC team and I had McDonnell, Boeing, GE, PAC, and Air Force inspectors. With the politics resolved, I started working with the PAC mechanics to get to know the aircraft and missile launch system and to set the standard for MAC at Holloman. My goals were to have the best team, fly more missions, and get the most interesting test programs. Boeing was flying the airplane and their pilot had the final say on safety. However, I had the final say on the Quail as well as the bomb bay configuration and test readiness.

One thing lacking from my education at Parks College was the ability to read electrical schematics. I asked Morris Laswell, the PAC lead, to pick his best electrician to teach me to read and use electrical drawings. He smiled and pointed out one of his mechanics, a muscular, friendly giant, Bob Vermillion. "There is your man, you ask him." Vermillion was well over six feet, six inches tall. The kind of guy you wanted on your side in a brawl. He squatted when he talked to people, and when he squatted to talk to me for the first time, the seams of his pants seemed about to burst. He was basically shy and nonplused with my request. He said, "I'll checkout some drawings tomorrow, and we'll work during the breaks." The tutoring sessions quickly expanded to the entire PAC crew. They were learning the systems, and I was learning the principles of reading schematics. We taught each other, sharing the knowledge which brought together every aspect of the mechanical, electrical, and electronic skills of my team.

Admitting I lacked skills and asking for help was the first unplanned step in building chemistry for the B-52 test team. On subsequent lunch breaks, the mechanics would take me aside to demonstrate their skills building wire bundles, waterproofing connectors, and writing good technical handovers for following shifts. I was surprised when asked to become a member of their Friday midnight bowling team. Joining their team was the final link in building the chemistry for our B-52 team.

Vermillion was one of the best teachers I have ever known, and Laswell taught me a lot about assigning work to make people stretch and grow. I started developing schematics for quick reference in building the preflight and flight checklists and as a troubleshooting aid. As a team, we learned the characteristics of the carrier aircraft, missile, and launch systems, and we could visualize the integrated aircraft system much like I would think three-dimensionally as a FAC in Korea. We conducted shop checkouts of the launch system on the floor of the Boeing facility and gradually developed the Boeing mechanics' confidence.

In short order, my PAC team was teaching all MAC and Boeing shop personnel the Quail launch systems. By the time I flew, I had hand-drawn electrical schematics of every installation placed in the bomb bay. From that day forward, I learned the space systems the hard way, color coding the integrated drawings, annotating with test data, and notes to find supplemental systems' data.

The flight test program started rolling, and without additional resources, I split my shop crew to cover twelve-hour operations, starting preflight at 4:00 a.m., engine start at 8:00 a.m., turnaround at 10:00 a.m., and a second flight around noon. The afternoon was spent on postflight data review and aircraft prep for the next day's activity. Interspersed with this was the perpetual battle to close the inspection paper of the three groups that had to buy off the airplane. Overnight I had moved into a leadership role, and without realizing it, I had adopted the best characteristics of my bosses in my work. I was fortunate that all had been great teachers.

I quickly fell in love with the B-52 and on three occasions, applied to reenter the Air Force, obtaining recommendations from both Boeing test pilots. It was the late 1950s, before we had made a serious commitment to the war in Vietnam and I learned that General Curtis LeMay, then Vice Chief of Staff of the Air Force, had no need for fighter pilots. Even today, the B-52s continue to operate, and I was glad when I heard that my 56-695, after a tour in Vietnam, found a home in the Hall Airpark at Tinker AFB near Oklahoma City. It was a special plane.

The Air Force was looking for a "Quick Load" package with four decoy missiles. My team and I were sent to Wichita, Kansas, for training on the new B-52 launch gear package. The training was timely, excellent, and complemented our self-study and provided a great chance to continue building my team.

We then established a time-integrated work sequence for the delivery of missiles from the McDonnell shop to the B-52 bomb bay. This process provided the procedures, responsibilities, shop and inspection requirements, transport mechanism, hazardous fuel loading, bomb bay installation, and power up and checkout of related B-52 systems. I believe that many of today's industry issues are related to product delivery delays because of the lack of an end-to-end structured process to ensure consistent quality and on-time production. Years later, at a quality management course at NASA, I found my belief stated by W. Edwards Deming, "If you can't describe what you are doing as a process, you don't know what you're doing."

***

Ralph Saylor XGAM 72 Project Manager, a great leader and teacher

In appearance, demeanor, style, and relations with his personnel, Saylor was a "roamer." Occasionally, I would look up and see him standing silently behind me or my team at aircraft sign off. I had the feeling that he was just itching to do the work rather than be the one in charge. He was six feet tall, lanky but physically imposing with a sun-bleached crew cut, a great bushy silver-blond mustache and eyebrows, and crystal blue eyes. In every conversation, his eyes carried the message to support his words. Leaders often adopt identifiers, which is something that immediately sets them apart from those they are leading. This is particularly important in civilian circumstances where there is no uniform and no markings to denote rank. Carroll had his colorful bow tie and suspenders. When I was in Korea, Marta made blue and gold scarves for my pilots. Later on, of course, I had my vests. Saylor's identifier was a tan "Aussie" hat and on cool mornings, a safari jacket. Driving an old beat-up Studebaker work truck with a gun rack, he was known by his team as the "Great White Hunter."

Saylor often encouraged us to exercise initiative. "Trust your judgment," he would say, "and never refuse to take on anything, regardless of your orders." This gave us license to exceed our "written responsibilities" and do what we believed was right. That could be dicey, because it might mean that you exercised authority on people you technically did not control and yet, Saylor had a saying for that too. "Dealing with people you do not control is easy so long as you cultivate them as friends and ask their help."

One of the most important lessons I learned from Saylor related to time. The White Sands Test Range covers over 11,000 square miles between El Paso and Albuquerque. Many tests are run in parallel and test range time is controlled in minutes from range entry to exit. "Time is an expendable resource," he would say. "It can be used well, ignored, or it can just slip away." We have all seen a football team in the "Red Zone" with the clock counting down, the game on the line, and the clock run out. Miscommunication, delay in the snap count, and confusion with the coach on the sidelines can all lead to failure because the clock waits for no one. Like the "Red Zone," when the clock runs out, there is not a second chance on the test range. Time is not renewable, and it is not recoverable.

Shortly after noon on December 18, 1959, I received a call from Boeing stating, "We have a problem, all power is lost to the missile, but the chase aircraft says that the missile is still in launch position." Normally, in this condition, we could jettison the entire launch assembly however wind tunnel testing had not been done so the circuit was disabled and the missile was hanging almost two feet below the B-52's landing gear. Landing on top of the missile was the only option. The B-52's tank vents and landing gear were aft of the bomb bay. The Quail engine start tank was on the underside and fueled with highly flammable ethylene oxide. When we landed, there would be one hell of a ball of flame. If the missile came off the launch shackles, it would hit the rear landing gear and blow the tires, causing major structural damage.

The B-52 circled while we discussed the situation and the SAC command post at Omaha came online. Boeing and I assembled teams, and the B-52 crew and chase aircraft confirmed the umbilical had separated. I described the rigging of the umbilical and clearly stated, "We are wasting our time, there is no way to reconnect the umbilical in flight! We should start working on things we can do."

I turned the discussion to other options with my PAC team. They suggested if we landed on the missile "softly," the pins in the carriage drive motor would shear allowing the missile to be pushed back up into the bomb bay. We had sheared the drive motor pins several times while learning to rig the launch track limit switches. The debate raged with Boeing holding out for other "yet to be defined" options. I convinced Saylor to push the land on top of the missile option and he got on the UHF radio with Al Perssons who was flying the plane. Perssons agreed this was the only option.

Discussions indicated there would be little difference between lake bed or concrete runway landings and all agreed that we should land on foam to smother the flame when the extremely flammable ethylene oxide start fuel torched off. Perssons the decision to land at Holloman and after a practice approach, came in for a "grease job" landing. After a brief flash of fire, the missile pushed up along the launch track, the B-52's landing gear touched down, and the aircraft continued rolling down the runway to a safe stop chased by an armada of fire and rescue vehicles. My PAC team's brainstorming of the aircraft systems, our development of procedures and schematics, coupled with the association we had developed with MAC engineering had paid off. Our team had our first flight test save.

***

In 2005 when I interviewed Saylor for this book, he began by telling me he was born in a "chicken house." Apparently, his dad was a tool-and-die maker by trade, but "He always had his hand in something else" so around the time Saylor's mother was pregnant with him, his father bought an old chicken farm in Edgefield, Ohio. The farm did not have a house, so he cleaned out the chicken coops and made those into their house. Back then, women tended to give birth at home and that is how, on March 20, 1923, Ralph Saylor came into this world in the same place where mother hens had laid eggs for generations.

In 1933, when Saylor was just ten years old, his father was killed in a road accident. He had bought a tanker to truck his own gas from the wholesaler when he came upon a beer truck parked on the side of the road. It being 1933, this had to be among the first beer shipments to Ohio after the repeal of prohibition by the Franklin Roosevelt administration. Sadly, the truck tailgate was open, the lighting was poor, the brakes on the tanker weak, so there was a collision and Saylor's dad was crushed to death.

Despite his early passing, Saylor had learned a lot from his dad. "He had a big influence on me because he never gave any direct orders, it was never 'do this, do that,' but 'Watch me' and 'Okay, now you try it.'" That was the way I learned." Listening to him, I recognized this was the way he had guided us at Holloman, too. He said that, while working he also learned respect for people who do physical labor because "You don't get anything done if you don't get your hands dirty and work hard." I thought of all those hard-working truck drivers who had given me those lessons in life.

Listening to him, I reflected on the fact that Harry Carroll, Ralph Saylor, Chris Kraft, and I all lost our fathers at an early age, during the formative years. When you lose your father, you mature at a faster pace and part of that is taking on responsibility earlier than someone who had the benefit of two parents.

Saylor joined the Marines in 1943. After Parris Island, he was transferred to Camp LeJeune where aptitude testing indicated he had skills in math and physics. Training took him through one year of highly compressed electrical engineering work at Wright Junior College in Chicago, followed by Utah State University, and finally the U.S. Naval Air Tactical Training center at Corpus Christi. After completing radar training, he was transferred to Cherry Point as the watch supervisor of the communications and radar section. Discharged in 1946, he worked in many capacities as a tech representative for Philco, where he developed aircraft radar and navigation systems, and for the Martin Company on the Matador Missile. All of this was before he moved to McDonnell in New Mexico to oversee the Quail Flight Test Program.

Saylor's Sunday School teacher kept a loose-leaf notebook with handwritten creeds. The students were encouraged to contribute to the book, look through the writings there, and use them as guidance. For Saylor, the creed he chose became his lifelong commitment:

I want to live my life up to my highest, fullest, and best. I want to treat all men with absolute equality. I will be as firm as justice demands. I will be heard, but I will also listen and observe and hope to learn good from others as I hope others will listen and observe and learn good from me.

***

John Parker was a MAC structural engineer who supported my team during Quail launch gear installations in the B-52. In June 1960 he was reassigned to support the McDonnell Project Mercury team at Cape Canaveral, Florida. In July, with the pending Quail test program completion, I was offered a flight test engineer assignment on the F4 Phantom II at Edwards AFB, California.

Discussions with Parker on the NASA Mercury Program convinced me that space was the future, and that an operational assignment with NASA would provide significant new opportunities. I responded to a NASA advertisement in *Aviation Week* magazine and in September 1960 I was offered, and accepted, a position supporting the Space Task Group (STG) at Langley AFB, Virginia.

EARLY SPACE

6

LEGENDS

Heroes get remembered, but legends never die.
                                    —*Babe Ruth*

Sometimes, when I think of leadership, I visualize the pyramids of ancient Egypt—mighty granite, limestone, and basalt structures built on strong foundations intended to exist for an eternity. When done right, leadership is similar, creating a foundation that is stable enough to outlast the people who built it. With organizations, this means creating a foundation centered on the qualities of trust, shared principles and values that will support complex, critical, and high-risk ventures. The Space Task Group was built from the ground up, on a foundation of leaders experienced in high-risk flight research that provided the stable platform of the manned space program.

In fact, I cannot overstress the importance of this, for a poorly constructed foundation will eventually give way and once it does, the only solution is to scrap it and start all over again. I may not have realized it back then, but this too was an aspect of competence. If we did not have the right people with the right knowledge establishing the foundations for space travel, we would never have achieved our goals.

I remember October 1957 when I returned to Korea from Taiwan and my crew chief, Sgt. Tom Gerdes, told me that the Russians had launched "a Sputnik."

I asked, "What is a Sputnik?"

Gerdes did not know much more than that. "It's a satellite that is circling the earth and beeping," he said, quizzically.

I shrugged my shoulders and kept walking. In fact, Sputnik was huge news as the first manmade satellite: a 184-pound polished metal ball measuring about two feet in diameter with several protruding radio antennae. Defense Secretary "Engine" Charlie Wilson (the nickname came from his days as head of General Motors) called it "a useless hunk of iron." He was wrong about the materials. Sputnik was made of aluminum and "useless?" Sure, but the mere fact that the Russians had beat us got under the nation's skin. Remember, this was still early in the Cold War and there was great concern among freedom-loving countries as to which side would gain technical superiority.

Even before Sputnik, the Russians were strutting their technical knowhow. In September 1949, American reconnaissance planes detected high levels of radioactivity demonstrating the Soviets had exploded their first atomic device much like the ones we had tested at Alamogordo, New Mexico. In 1953, less than a year after we tested our first hydrogen bomb, the Russians tested one too. At roughly the same time, the F-86 Sabre and the Russian MiG-15 were competing over the Korean Peninsula. Early on, the MiG-15 appeared the superior aircraft. John Boyd, a FWS instructor, was puzzled by the Sabre's marked superiority because the two aircraft seemed evenly matched on paper and did not quite add up to the 10-1 kill ratio. After more analysis, Boyd concluded the Sabre possessed a quicker instantaneous rate of turn allowing it to more effectively transition between maneuvers and giving the pilots an advantage over the MiG. The American pilots also had better training. MiG testing indicated that above .92 Mach, the elevator response was limited. Test pilot Chuck Yeager described it as, "A quirky aircraft that's killed a lot of pilots."

The Soviets followed their first Sputnik launch with Sputnik 2 carrying a dog named Laika, the first animal to orbit the earth. Just three years later, while driving to Langley Air Force Base, Virginia, to join the STG, I was only beginning to recognize what this meant—a competition with Russia for superiority above the earth's atmosphere that would dominate my life and that of over 400,000 American workers for a decade.

With no internet, CNN, or Fox, and meager offerings at the small library at Holloman AFB, I had had no source for information on NASA or Project Mercury. The only news I could find was in *Aviation Week*, a magazine which began publishing in 1916, but changed its name in 1958 to *Aviation Week Including Space Technology*. That alone spoke to the coming changes.

Still, reporting to my new position at NASA at the age of twenty-seven, I arrived prepared. I was well-schooled in aviation and flight test by men like Harry Carroll and Ralph Saylor, members of the so-called "Greatest Generation"— NBC journalist Tom Brokaw's honorific for those who came of age during the Great Depression and then served in World War II. The young men and women of that time faced extraordinary challenges including economic catastrophe and a world at war but emerged victorious largely because they employed a unique combination of humility and courage. I could see those qualities in Carroll and Saylor, and I benefited from their demeanor. They prepared me not for a heroic life, for only a fool sets out to be a hero, but instead, a life of duty, honor, and service.

***

President Eisenhower signed the National Aeronautics and Space Act on July 29, 1958. This would lead to the creation of the National Aeronautics and Space Administration (NASA) on October 1, 1958. In November, the Space Task Group (STG) was established to manage America's human space flight programs with staffing from the Langley and Cleveland Lewis Research Centers. Within months, it was augmented by thirty-two Canadians from the Avro Arrow flight test program. My assignment to the STG meant that I served under a legendary leader from that era, Walter C. Williams.

Walt was a big bear of a man who grew up in New Orleans and earned a degree in aeronautical engineering from Louisiana State in 1939. A year later, he joined the National Advisory Committee for Aeronautics, or NACA. Established in 1915 (both Orville Wright and Charles Lindbergh were early board members), it evolved into the nation's prime research organization on the future of flight. NACA proved its value during World War II when superiority in the air was essential to victory.

Space Task Group (STG) leader Walter C. Williams

Towards that effort, Williams worked on the design of fighter aircraft like the P-47 Thunderbolt, P-51 Mustang, and the Grumman F6F Hellcat, before moving post-war to the Bell X-1 planes and succeeding generations of aircraft and fighters powered by rocket engines. Combat veterans like Gabreski, Ken Chilstrom, and Chuck Yeager were put to work testing the new planes. On October 14, 1947, Yeager famously broke the sound barrier flying an X-1. All the younger aviators revered Yeager and around the STG, Williams by association received recognition for involvement in that historic flight.

Like my prior bosses, Williams was a hands-on leader and I considered it a privilege to work as a member of his organization. His comprehensive grasp of things intimidated many people. He was notably brusque in communications. One day during a test count that was not going well, he told Mercury Flight Director Kraft, "Call a hold! I'm going to the pad to sort those SOBs out!" The toughest part was simply keeping up with him. During briefings on mission rules and test planning, I would find he was always ahead of me. He corralled the astronauts, NASA contractors, design and test teams, the media, and the politicians, keeping everyone in line. Plus, late at night you could regularly find him in the bar at the Holiday Inn, still working the problems. To Williams, there was no beginning or end to the workday, it was one continuous loop. The man just never seemed to sleep.

I remember in November 1961, while preparing for MA-5 (the Mercury launch which took the chimpanzee Enos for a ride), I was assigned to brief Williams on the mission rules. This was my first time working directly with him, so naturally I approached the assignment with some trepidation. I encountered a man with the build of a barroom bouncer and the manners to match. As I spoke to him, he was bathed in a blue cloud of cigarette smoke. *God help me if I cannot answer his questions*, I thought. Then, about ten minutes into my briefing, I had the distinct impression Williams had fallen asleep. Not knowing what to do, I continued the briefing. Then I noticed Williams moving. His eyes were still closed, but one hand snaked its way to his desktop and shook out a couple round Necco Mints from their box. Williams tossed them into his mouth, finished the mints in a few bites and then seemingly went back to sleep. When I finished, his eyelids parted and he sat bolt upright. He then proceeded to summarize every point I made and said, "Let's do it!" If a beer was handy, I would have celebrated.

We reported to Williams, but Williams reported to the even more legendary Robert Gilruth whose association with NACA and the Langley Research Center dated back to 1936. Gilruth has been referred to as "the father" of the manned space flight program and I think that is an apt honorific. He was a big man with a bald head, a laconic demeanor, and a love for pipe smoking. Gilruth grew up in Nashwauk, Minnesota, where he made model airplanes powered by rubber bands and entered them into competitions. After reading about NACA in the pages of *The Saturday Evening Post*, he wrote to request NACA's research on airfoils which he applied to his model planes. After high school, he went to the University of Minnesota where he eventually earned a master's in aeronautical engineering. Like Williams, he worked on supersonic flight research in the 1940s and 1950s, heading up NACA's Pilotless Aircraft Research Division (PARC). However, his recent work involved satellites, not planes, making him perfectly suited for the next frontier.

By November 1958, Gilruth's plans for a manned satellite program included establishing the STG and Project Mercury. Plans were moving quickly to bring together America's many disparate space initiatives under one roof. Indeed, it replaced NACA, and incorporated the Langley Research Center, the Ames Aeronautical Laboratory, and the Lewis Flight Propulsion Laboratory into the NASA organization. NASA now controlled the Army's missile program, which included the efforts of German rocket scientist Wernher von Braun, the Air Force ballistic missile program, and the Naval Research Laboratory. A key factor in the creation of NASA was the need to expand the nascent space program from its initial military purposes into civilian applications. Also, the new agency needed to stop the interservice competitions that too often pitted the Navy against the Air Force and the Air Force against the Army. We already had a formidable rival in the Soviet Union. NASA forced us to join in a common purpose.

While Gilruth and others ensured the STG was well-prepared for the technical challenges, it lacked personnel for the financial, budgeting, and management of the developing space effort, and therefore addressed these issues on an ad hoc basis. As a result, administrative issues plagued NASA's early years. Putting a good face on it, Gilruth said, "The fluidity of the organization allowed it to do things and could only occur in a young organization that had not yet solidified its functions and prerogatives,"[7] but he was right. Gilruth and the other pioneers were building that critical base foundation of the "pyramid," one intended to outlast all of them and that meant establishing the organization's values, culture, and purpose.

Flight Test was the common fabric of the STG. Besides Williams and Gilruth, we had Sigurd Sjoberg, project engineer for the D-558 Skyrocket that Scott Crossfield flew in 1953 at Mach 2, twice the speed of sound. It was the first aircraft to accomplish this feat. We also had Jim Chamberlin and John Hodge who were recruited from Avro of Canada where they worked on the CF-105 Arrow Interceptor, the world's top performing aircraft at that time. In an act of supreme shortsightedness, the Canadian government had cancelled the program, leading to the termination of the engineers and STG became the beneficiary. Jim Chamberlain was a gifted Canadian aerodynamicist who worked closely with Dr. Gilruth on the Arrow. Chamberlain was assigned as the Chief of Project Mercury engineering. He was tall, thin, nearsighted, and wore round glasses. Despite being nearsighted, he was an excellent swimmer and water-skier.

One early evening at the motel at Cape Canaveral, we received a call to go search for Jim. He was missing after taking a fall while water-skiing on the Banana River. We got all the motel flashlights, then bought more at the hardware store and stood along the causeways as the tide was going out. The search continued until early the following morning when Jim was found hitchhiking on the highway. He had spent the night on one of the little river islands before swimming to the mainland.

The members of the STG provided a broad spectrum of skills. Like me, some were former military pilots, others were young engineers from the satellite programs, and even young college graduates experienced in the new space technologies. Just looking at the people, you felt they were radically diverse. In the military, everyone was similar, cut from a common stencil that only allowed minute variations. Everyone in the STG was different. There was a veritable rainbow of personalities, mannerisms, and speech. They were friendly, almost collegial, and seemed to be dreamers dealing in ideas rather than events. The result was a government organization composed of equal parts experience and youth that was supported by a healthy, competitive aircraft industry engrossed in restructuring because of new and exciting technologies.

We were entering a new age. The turbojet engine was replacing the piston engine, stimulating intercontinental air travel. Intercontinental ballistic missiles (ICBMs) were replacing the massive fleets of aircraft used for defensive and offensive missions, and the launch of Sputnik expanded the Cold War into space. Were we behind the Soviets? Yes, but spurred by Congress and the press, NASA would now assume the role of restoring confidence in America as a leader of the free world. The leadership foundation was established.

As the STG grew from an organization of less than a few hundred to a staff of thousands, the very close working and personal relationships that developed in the early years never diminished as we moved into Projects Gemini, Apollo, Skylab and even the early Shuttle. Our communications, whether technical, financial, or risk-related, were two-way exchanges that recognized the risk of our work and the need for timely decisions. The associations developed in our early years were surprisingly personal. Many times over the years I was asked to brief the NASA Administrator or a Program Manager on a personal basis or even receive a phone call from Washington to address a shuttle software question with a CNN reporter.

During mission operations, I often had coffee with the Program Managers and astronauts to address issues prior to the daily program meeting. On occasion, they sat with me on the console at night or during the first lunar landing. The relationships developed in our first decade would carry us through the times of challenge, triumph, and tragedy of our early space years.

7

CHARACTER

Character is destiny.

—Heraclitus

STG pioneers and teachers Chris Kraft and John Hodge

Whatever role a person chooses in life, character defines their destiny. Character, the sum of an individual's underlying traits or, what I call, "substance"; the attributes and behaviors that underpin a course in life. Character is what an individual stands for. In my writing, you have learned of Harry Carroll, Ralph Saylor, Mister Coleman, and Wendell Dobbs. All of these individuals typified integrity, perseverance, and resilience during a productive and successful life.

They taught me leadership and of "IT!," the team chemistry for success.

I encountered others cut from the same cloth. In particular, Chris Kraft and John Hodge. I was an early member of the STG and had the good fortune to work alongside Kraft and Hodge in establishing the organizational culture of Mission Control and addressing the challenges of our work. Leadership is the ability to exhibit toughness of purpose while best utilizing the talent, energy, and abilities of a group to competently proceed toward an objective. The power of the leader comes from their character—who they are, what they stand for, and their ability to mold an organization in their image. Kraft and Hodge were initially assigned as assistants to Mathews, the Chief of Flight Operations, and I was assigned to the Flight Control Operations Section along with several Avro engineers. Kraft and Hodge both capitalized on the lack of a formal structure to recruit individuals and assign them to accomplish work they believed critical. They both preached Saylor's philosophy to me, "Trust your judgment and never refuse to take on anything, regardless of your job description." In the beginning, with limited top-level oversight, what they wrote and did became policy. While the environment of space flight was vastly different, the mitigation of risk through planning, risk judgment, and experienced hands on leadership was similar to aircraft test and could be readily adapted. When the STG was formed, the Mission Analysis Branch (MAB) was established as NASA's first formal manned spaceflight mission planning organization. A few specialists had experience in orbital mechanics and the MAB was tasked to define what mission planning work was needed to support the test flights and the manned missions. The MAB was immediately recruited by Kraft as a key element of his growing organization.

***

Chris Kraft was born in 1924 in Phoebus, Virginia, and John Hodge in Leigh-on-Sea, Essex, England in 1929. Both were trained engineers. Kraft attended Virginia Polytechnic during World War II while Hodge was evacuated from London during the Nazi London Blitz and, at the war's end, received his engineering degree from Northampton College of the University of London. With our shared interests in aviation, flight test experience, and similar ages, we established a working association: a kinship in direction and solutions that seemed at times intuitive. At that time, unknown to me, we three were building the core of operations for manned spaceflight for Mercury and the future manned space programs.

When I began at NASA on October 17, 1960, I was initially assigned to Goddard but by a stroke of good luck, I was quickly reassigned to the Langley Operations Division by Kraft for Operations in Mercury Control. The assignment provided my first taste of his management style: no discussion, no work history, not even time for a question, just do it! To date, all my experience was reviewing Project Mercury and IBM network familiarization manuals and reading Paul Johnson's text on Mercury Control. This was the first document providing working descriptions of the controllers and the Mercury Control team structure.

I was used to taking orders, just stepping up to the plate and getting to work, but this was different. In all prior work, I had time to get my feet on the ground. This assignment was something that had never been done before. The Cape assignment was the first of my "learn by doing" experiences and it was key since it provided my introduction to the Cape Mercury launch team and Mercury Control. The day I walked into the Mercury Control room, I had the sensation I had found my life's work. Because of the limited workforce, I ended up working as Kraft's scribe for Mercury Control, Goddard communications and the tracking network, providing a good perspective of the key personnel as well as the total team.

The STG was comprised of three divisions: Flight Systems, Engineering, and Operations. I was assigned to Operations. Charles Mathews was the Operations Division Chief and Chris Kraft was the Assistant Chief. Kraft had a counterpart at the Cape supporting launch operations.

While there were still many empty offices, you could feel the energy build in short bursts of infrequent conversations, the briskness of everyone's movements, and the evident preoccupation of those in the hallways.

The Philco Tech Reps were the teachers for the young remote site teams.

I was in an office with Navy Captain Paul Havenstein, John Hodge, and Sig Sjoberg, Kraft's assistant. The Operations Division structure was simple; everyone was either a chief or an office head with few employees. My relationship with Sjoberg and Hodge was one of learning by listening and our common background in flight test and the co-location in the office rapidly fostered a close friendship and level of trust. Sjoberg's quiet but firm demeanor belied his vast flight test experience in rocket-powered aircraft and his selection to evaluate several European fighter aircraft for NATO.

The team I joined under Kraft consisted of sixteen Philco tech reps and about fifteen young NASA engineers that I considered "hired hands," drafted for mission support from the four branches of Mathews's Operations Division. While many tried, operations was often not their "cup of tea." The flight surgeons assigned to the tracking stations were military personnel from the nearest DOD facility. In many ways, the surgeons and the Philco tech reps took the "hired hands" under their wings, rapidly developing their skills to the time-critical, high-risk environment of NASA tracking stations which were scattered around the globe. Kraft did not give me "instructions" because he did not have any. My first critical lesson working for Kraft in the STG was that we were "learning by doing." The "Procedure's Console" and "Flight Director's Console" were adjacent, so I elected to perform duties as the "Procedures Officer" supporting Flight Director Kraft. That decision placed me at the center of all message traffic during a mission.

My second action was to quickly get to know the Philco Tech reps and the few NASA engineers assigned for remote-site duty. The eleven tracking stations and two ships were essentially mini-control centers. My office proximity with the personnel provided experience to select the teams for site assignment. Voice and data capabilities were limited, so teletype messages provided most of the communications links for Mission Control.

Like so many in the space program, Kraft initially worked for NACA at Langley. He was a close associate of Williams and in his work leading up to the STG, had done research on gust alleviation (dampening the effects of rough air on an airplane, thus reducing turbulence) and on the flying qualities of the Navy's F8U jet fighter. One of the pilots he sparred with on that project was a young Marine named John Glenn. Chris expressed his opinions well and had three modes. He did not like equivocation. If a person were wrong in his opinion, he would let them know. Often, he would go into his "If you're a good manager, you will find out there is more than one road to Dallas." Other times if he was unsure, he would go into teacher mode, asking questions, listening, or providing input. Like a good fisherman, he knew when it was time to sink the hook or retrieve the bait. Kraft had a leader's stature and demeanor, a class act probably derived from his Virginia Polytechnic training while a member of the Corps of Cadets.

When I arrived at Langley, my clothing consisted of military uniforms and that of the Holloman flight line. I did not own a suit. Kraft did not waste words. He was succinct, "What you wear is how you present yourself to the world." While the astronauts wore short-sleeved Banlon's (a type of polo shirt), I wore a shirt, tie, and jacket. Generations later in Mission Control, I would repeat the same words to my team. For the next two decades, the relationship with Kraft would dominate my professional life. Many of his personnel thought of him as a teacher. Several considered him a leader. I played Kodokan Judo with Dutch von Ehrenfried, John Llewellyn, Bill Moon, and several other early controllers. Judo taught respect for those who mastered the sport. I considered Kraft a Master at leadership.

In the early years, Chris became a friend, a role model, a mentor, and a teacher. No one schooled me longer or better in the art of leadership than Chris. I initially treated him as my "senior officer" because of the high level of professionalism he applied to every aspect of his work. We formed a bond that worked well for years. Similar to my relationship with my flight instructor, Jack Coleman, initially I could only address him as "Mister Kraft." After eating barbecue, playing volleyball, and being tested together by our work, he became "Chris," or simply, "Flight."

***

Over the years, I learned a lot about him, especially during Gemini when we shared an apartment at the Cape. Later, when I interviewed him for this book, I received an even deeper sense of the man. Leadership qualities emerged early for Chris Kraft.

"When I was a kid, I was the guy that got on my bike and rode around the whole town of Phoebus,

Virginia, to pull together a baseball team. Just about every place I went, I seemed to be able to tell people what to do. When I was at Virginia Tech, I had already been in the drum and bugle corps and knew how to parade and so I was the natural guy to teach people how to drill."

When the war came and the older cadets were called into service, it was left to the younger Kraft to adopt an even firmer hand on the levers of leadership. "Whether you like it or not, 200-250 guys depend on you to be the leader." Called to duty, he responded. Kraft's normal voice was commanding; he spoke few words, but his commands and intentions were always clear. Often, his directions ended with a question mark, implying the direction he thought we should go unless we raised a question.

Like many I have already described, you can find the seeds of leadership in Kraft's upbringing. Harry Carroll, Ralph Saylor, and I lost our fathers when we were young. So, too, did Kraft. Chris's father, who suffered from depression, did not die while he was a boy. Still, Chris lost him in that his influence was, "More shadowy than substantive." When interviewed, he said that because of depression, his father was ". . . almost a zero in my life, unfortunately. I accepted life as it was without wondering if it should be different." But in our interview, he said his father's failures ". . . forced him into making his own decisions early in life." Early life forced a maturity beyond all our years.

***

While Kraft was building his team for the Mercury Control, Hodge was also building his team for the Bermuda Tracking Station. Mercury missions were launched on a trajectory that passed over Bermuda. Mercury Control was responsible for the overall conduct of the mission, however, as the Atlas rocket approached orbital cutoff, the Cape would lose contact shortly after achieving orbit and operations would be handed over to Bermuda. During this critical period, Hodge's Bermuda Flight Dynamics Officer (FIDO) Glynn Lunney provided backup support for the orbital "Go/No Go" decisions and, in case of an abort, provided tracking predictions for abort splashdown. To protect against a voice communication loss during launch, I dictated launch phase events to a teletype operator in Mercury Control. The words were sent in "real time" to an operator in Bermuda who then read my words to Hodge's team throughout the launch period. The relationship of these two sites tied me closely to both Kraft and Hodge throughout the Mercury Program.

While I worked with Kraft most of the time, I was assigned organizationally to Hodge for almost a decade. Kraft was one to go straight at a problem, attack it and overwhelm it. Hodge studied approaches, encircled them to see what developed, then figured out how to turn the outcome to your advantage. John and I had both worked with more advanced technology in our aircraft programs than that being used during Mercury. The Mercury challenge, however, was to flight test a spacecraft at 17,000 mph with a teletype system that dated back to the days of the Pony Express. The voice communications system resembled that used by ham radio amateurs relaying communications through several sites to the tracking stations. With no computer onboard the capsule, Mercury Control confirmed the capsule attained orbit and then periodically provided calls to the astronauts advising them how long and how many orbits they could remain in space, "You are Go for 1 (or 2 or 3) orbits" and with additional tracking we advised "Go for the mission."

***

John Hodge was an Englishman and to me, instantly likeable. His spoken voice, demeanor, and dress set him apart from the others in the office. His wool, tweed jacket with leather patches on the elbows coupled with his graying hair and pipe made his appearance distinctive as well. To him, a pipe in his hand was an instrument for speaking and making a point. When they arrived, the Avro engineers formed twenty-five percent of the Space Task Group at Langley. The merger set up an interesting situation. The Avro personnel experience was test-oriented and production-oriented so the merger with the research-oriented Langley personnel was a very good meld. Hodge described it as a good fit, "There were plenty of jobs, so everybody sort of looked around and said, 'That needs to be done,' and you talked to somebody, and they said, 'We'll do it,' and that was it, and you went out and did it."

I would classify Hodge as a sagacious personality, with wide experience and learning. Postcollege, we had both worked in factories, had flown, and it provided a trait where there was often a niggling feeling, and therefore a withholding of direction until we had given it more thought. This thorough characteristic was respected within the team environment. He did not micromanage, and we trusted his judgment. If he wanted to delay action, we considered it in the team's best interest. Every team draws benefit from a sage who says, "Let's just think about it a bit longer."

My own engineering, data reduction, and flight experience came into play when I invented the Procedures role, wrote countdowns, and mission rules. I made many phone calls, often with stupid questions, which got good answers. Everyone, *everyone*, was incredibly helpful. Soon, during the testing with the launch team, I would be listening to people on the other end of the communications loop who I now knew as a friend, and I knew their parts in the enterprise. There was an acute need for experience and having people around who had "been there" and "done that" provided the foundation for effective working relationships.

When I worked with John, I always let him provide the opening words. He would then take a few puffs on the pipe, set it aside, and it was time to talk. John was a teacher, one from whom I would learn about the global perspective of our effort in space. He was a constant challenge with his questions, always seeking opinions on direction, the next step, and on the many new employees entering our organization. His philosophy of operations was to, "Look for simple answers but question everything." This was good guidance as we ventured into the new arena of spaceflight.

***

In September 1940, when the United Kingdom did not pursue peace after Germany's victory over France, an all-out air attack on London and other major cities began. The Germans aimed to weaken the morale of the British people and force the country to surrender. John, like most English children, was moved to the countryside with his mother, to a safer place. The area they selected was near Braintree in Essex. However, it was a poor choice as it was a corridor for the German V-1 buzz bombs. His father came and relocated them to the North London area, which he considered safer.

The British school system is significantly different than the system in the United States. Children often entered high school at age eleven and graduated at sixteen. Prior to entering school, a screening examination established their future potential, leading to either a university or a vocational school system. John was qualified for the university system. Normally at that time, only six percent of the British population ever entered college. Graduating from high school at fifteen, Hodge planned for a future as a biochemist. However, military personnel were assigned a higher college priority so John settled for an engineering education, graduating at twenty. His father insisted he should "Knock off the square corners and work in an aircraft factory to find out how ordinary people live." In retrospect, John considered it "the finest education he ever had." The factory work taught him the "stupid things engineers do in design." He considered this hands-on learning experience key to his future as an engineer. He spent four years in the University Squadron, which was similar to the American ROTC. This provided him with flight training in a Tiger Moth, a tremendous biplane, which was easy to fly but difficult to fly well. He then graduated from North Hampton Engineering College, located in the middle of London, in 1949.

John had been dating a nurse who put off marriage, indicating she was going to Canada first but John had a different view. He and Audrey were married and left for Canada six months later. It was his good fortune to find employment with Avro of Canada which was producing the CF-100 aircraft for use by the Canadian Air Force as a jet trainer. In 1956, Avro initiated development of an extremely high-performance aircraft capable of intercepting Russian bombers coming over the polar region. John, twenty-five-years-old, initially worked in the design of the intake, engine installation, and exhaust system using NACA test data.

Then, he was assigned to the Loads Group and eventually to the CF-105 Arrow flight test program. Three years later in the fall of 1958, an Arrow aircraft was on a speed run to the edge of Mach 2 when the radar operator who could no longer suppress himself, commented, "Will you look at that sonofabitch GO!" John was the flight test engineer on the number 5 aircraft with the more powerful Iroquois engine and testing was expected to exceed Mach 2.

CF-105 Avro Arrow - The proving ground for high speed flight

Unexpectedly, while anticipating this critical flight test on Black Friday, February 20, 1959 the Avro workforce was advised by the company president, Crawford Gordon, that the Avro aero project was being cancelled by the government and the factory would close. "Everyone except Security was told to go home, take your stuff with you, and, when we have more information, we will call you."[8]

In one stroke, 14,528 employees of Avro Aircraft and Orenda Engines were laid off. Three days later, all employees were advised to come in and clean out their work areas.

A month later, the Canadian government arrived with welding torches and destroyed every aircraft, including the drawings and tooling, and sold virtually all the pieces for scrap. "It was terrible to go down to the production line and watch them cut those airplanes up," John said. In fact, there were ten airplanes destroyed as well as the ones in flight test. The aircraft with the Iroquois engine was scheduled to fly and exceed all current aircraft speed records within just a few days of this event. Only the nose section of the sixth aircraft and other small parts remain.

Jim Chamberlin, the chief designer at Avro, had been working with NASA Lewis on the propulsion system. At the same time, NASA personnel indicated they were looking for people to join the STG. Chamberlin met Robert Gilruth at Langley and they reviewed the STG organization charts and available positions. Chamberlin then returned to Toronto to prepare resumes for 150 new hires and transmitted them to Chuck Matthews and Max Faget. After the review, the STG leadership selected fifty to be interviewed. The following day, they offered civil service positions to thirty-five people. Hodge was offered a GS-14 position and while at the time he didn't know what a GS-14 was, for a thirty-year-old, it turned out pretty well. The Canadian contingent arrived the same week as the Mercury 7 astronauts were announced. Hodge was the softest spoken of the group and once you got to know him, behind his words was a hidden strength. In preparation for the Gemini Program, Hodge and I would be assigned duties as Flight Directors.

8

LEARNING AND DOING

The task of the leader is to get his people from where they are to where they have not been.
—Henry Kissinger

There was no precedent for the manned space program. As NASA labored to get itself organized, it realized it would inherently be markedly different from NACA. While the agency would continue to conduct research, it also acquired operational responsibilities for the systems it would design, build, and test. Much of this work would be accomplished by issuing contracts. Operations needed to be built from scratch, and I was one of the lucky ones who would build it. The environment of the STG was reminiscent of my first weeks on the B-52 at Holloman where Saylor was building his team, making initial assignments, and then turning us loose. Saylor however had a workforce with aviation and flight test experience, all that was needed was for someone to take charge and Saylor did.

This was not the condition when operations were initiated in the STG. Phone books provided the only organizational structure, there was limited program documentation, and there was no top-level experience and leadership on space. The program merely existed in the minds of the senior engineers on the STG, on the drawing boards, and at McDonnell. The operational leaders, Kraft and Hodge, had no "direct reports." They just derived their support from those close at hand or someone they found in the halls. Their approach was simple; just walk up and assign an action or say, "I need some help." There was no visible STG recruiting organization assigned to feed personnel to flight control operations. It seemed most of the new hires went to engineering and I believe it was probably because "flight control and operations" was a job still waiting to be defined.

I did not grasp the scope of work until two weeks after arrival when Kraft sent me to the Cape with the assignment to write a countdown and mission rules for the first Mercury Redstone Launch. There, I found many who were searching like myself. In retrospect, this was a startup organization for a first-of-a-kind team and our job was to invent manned space operations.

Mercury Control was a room like "King 1" at the Holloman Range Control Center, equipped with radar plot boards, wall clocks, desks, and consoles with communications panels. In Mercury Control, there was a massive wall display map with side-plotting screens like those aboard ships in WWII Navy movies. Technicians wearing headsets—some standing, others seated in front of communications panels with blinking lights—and a muted hum of voices filled the room. I had seen this before; my past and present were converging. In the early days of space, I learned to trust my inner voice.

> **When there is no rulebook, the only remaining check and balance is your inner voice, and it is important to trust that voice.**

I immediately felt at home with my new job. I had worked in a similar control room before, only this time it was with rockets and capsules instead of missiles and aircraft. I had a sense of destiny and knew I would again build a team from scratch to reach a superior performance level as I did at Holloman. I introduced myself to the technicians and was soon making the rounds of the telemetry, display, communications, and tracking areas that I knew so well but on a smaller scale. John Hatcher, the Mercury Control Center facility coordinator, briefed me on the Mercury Control interface with range control, blockhouse, launchpad, and Hangar S where the Mercury capsule and astronauts were located. The new part was when Andy Anderson and "Esh" Eshelman took me to the teletype area and briefed me on communications and the message flow coming from tracking stations. Esh surprised me with the directive, "All message forms will be printed with #2H lead pencils." I thought that if that was the directive, I would not buck the system. My new job felt like my old job but one of high energy supported by solid, hands-on people in love with their work and with the ever-present clocks counting down to events.

Refreshed and returning to Langley, it was time to get to know the controller personnel assigned to operations. On paper, my boss was a Royal Canadian Air Force pilot and flight test engineer, Fred Mathews. He was a Clark Gable-look-alike with an interesting flight test background. However, my time with him was brief because Chris Kraft commandeered me for his Mercury Control team. Half of Kraft's civil service workforce, the CapComs (staff who direct the site and communicate with the crew in the spacecraft), spent most of their time performing their primary engineering duties in other divisions instead of preparing for missions. Sixteen Philco corporation tech reps constituted Kraft's team for the operations at the thirteen Mercury tracking stations. Like my PAC mechanics at Holloman, they were older and well experienced in early satellite operations from their time at the Vandenburg, California; Kodiak, Alaska; and Kwajalein Atoll, Marshall Islands tracking stations. Introducing myself, I found they had a wealth of knowledge on satellites and since they had no other assigned leader, I told them they were working for me.

***

One month and four days after joining NASA, on November 21, 1960, I was seated in Mercury Control for the unmanned Mercury Redstone 1 (MR-1) launch. My console was to the left of Flight Director Chris Kraft. There were two trajectory controllers, Carl Huss and Tec Roberts, that I had yet to meet, seated in front of scribing plot boards and an assortment of technicians. There were no engineers monitoring the dummy spacecraft atop the rocket, and for the size of the Mercury Control room it felt very empty.

Unmanned Mercury Redstone 1 (MR-1) launch NASA's "Four-Inch Flight"

When the firing command was sent Kraft's console television picture showed smoke on the launchpad through which we presumed the rocket rose. The television camera tracked upward, hesitated, then looked down at the launchpad. After the smoke cleared, the Redstone rocket was still on the launchpad. We had launched the Mercury Capsule escape tower and we did not know what to do. We sat in Mercury Control listening to the launch team debating options to safe the Redstone while waiting for battery power to run out and the Redstone to return to a "safe" condition. It became obvious that Kraft's team, as well as the launch team, had limited knowledge on both the rocket and spacecraft. This was an inauspicious beginning, but one from which we recovered quite quickly. One month after our first Redstone launch attempt, we returned to the Cape, closing out 1960 with MR-1A our first successful Mercury Redstone launch.

STG Pioneer Engineer Arnold "Arnie" Aldrich - Arnie never stopped growing, taught by example, and became the "Competence" model for the Mission Control teams.

With the first launch behind me, I began building a team as I had at Holloman. After describing the impact of the first Redstone launch attempt, I sat down and asked the Philco tech reps to teach me about remote site operations and the functional tasks to manage the tracking stations when the capsule was in contact over the site. On occasion, two other engineers joined the meetings, Arnie Aldrich and Paul Johnson. They were tasked with writing a Mercury Console handbook, and were seeking details to further refine the controller's handbook they were developing.

This was my first introduction to Aldrich, a young electrical engineer who already had a good grasp of operations. He joined the STG a year before me and carried a wealth of information on the tracking stations and network. He was a good listener, what we called a "soak," and asked solid questions. On paper, we were both assigned to the same section; however, we seldom crossed paths. He was assigned as a Mercury CapCom and told me he was going to the Canary Islands station to develop remote-site procedures with Paul Johnson from Western Electric corporation. I quickly recognized Aldrich as a "doer"; he did not ask permission. If there was something to be done, he just went and did it. For the orbital missions the teams staffing the Guaymas station in Northwest Mexico were required to meet with the American ambassador prior to continuing to the tracking station. When Aldrich's team arrived in Mexico City, for the orbital MA-5 mission with "Enos" the chimp the ambassador was having a party and Aldrich was told to wait until the next day. Aldrich, unwilling to delay the 950-mile trip elected to take his team to the site.

Months later, recognizing we were woefully ignorant of the Mercury capsule systems, I obtained Kraft's approval to assign Arnie to develop a Capsule System Operations Handbook. He would work with Ed Neiman and Dana Boatman, two McDonnell engineers from the Saint Louis plant. The first Flight Controller Handbook for Capsule 13 was produced by Aldrich and his team on the December 1, 1961. The "Learning by Doing" used to develop this handbook was the first major step in developing the flight control team technical skills and chemistry, providing the foundation for "IT!"

Team chemistry, IT! is composed of task and social components that elevate, performance by igniting the inherent capabilities within the team. "Learning by doing" provided the foundation of the task chemistry and it fostered the exchange of knowledge with other team members developing the social chemistry.

To fully appreciate the Mercury lessons, you need to step back in time to the early 1960s. Global change was common, the European colonial period in Africa had ended, and anti-American nationalism was common in Central America and Africa. Commercial jet transportation was in its infancy and computer technology was limited to laboratories. With no satellites, global communications consisted of landlines and underseas cables to Europe, Hawaii, and Australia. Radio and teletype links provided connections from the cable terminals to the ship locations and the stations in Africa and the Pacific Ocean. At many remote sites, the Go/No Go decisions were in the hands of engineering personnel "borrowed" from other STG organizations who had limited experience in their critical mission role.

The suborbital Alan Shepard flight on May 5, 1961, while of short duration established America's Man in Space Program, providing a much-needed boost for the young space team. It was soon followed by a second suborbital flight with Gus Grissom that cleared the way for a manned orbital flight in 1962. The evaluation of controller performance during the orbital Mercury Atlas 4 and 5 in late 1961 started the shakedown of the mission manning. The Procedures position in Mercury Control was the voice and teletype communications node during missions and quickly provided the perspective on individual controller and team performance at the tracking stations. The first missions identified the few controllers with the necessary initiative to effectively support the mission work and provide the leadership for the contractor and civil service personnel. Kraft had a good memory for the controllers at the remote sites who had performed well and as we moved into the manned orbital missions, we located the best performing controllers at key sites. Midway in the program, Aldrich, Llewellyn, and Lunney were moved from the remote sites into Mercury Control positions.

The newly established Gemini Program was requiring support and Kraft, now Chief of the Flight Operations Division, began relocating his organization to Houston after the MA-6 Glenn mission. At times I felt I was back in the Air Force rapidly moving between assignments in Virginia, the Cape, and Houston

I was well behind the power curve in all aspects of mission preparation when help arrived. Manfred "Dutch" von Ehrenfried, a physics and math graduate from the University of Richmond reported for duty. I virtually unloaded on my new hire giving him console procedures, mission rules, and the need to become adept in communications command and control. Dutch was a quick study and soon filled in for me on my console duties. He was also assigned the duty to find "affordable" housing for the controllers moving to Houston.

Kranz, Kraft, and Williams during Mercury Atlas 6

The February 1962 Mercury Atlas 6 orbital mission with John Glenn provided critical lessons in further defining the mission command and control functions and responsibilities.

Early in the mission, telemetry indicated the heat shield, critical for reentry heat protection, may have come loose. There was no telemetry redundancy, and no readily available engineering data on the switch location and rigging, so the issue was, could we trust the telemetry for a life-critical decision? Kraft was surrounded at his console by Operations Director Walt Williams, Engineering Chief Max Faget, and McDonnell Vice President John Yardley, all debating retaining the retrorocket package, after retrofire, to hold the heat shield in place. The debate on reentry options continued for almost three hours. Flight Director Kraft believed he had the decision authority, stating he believed the telemetry was in error and reentry should proceed normally. Williams indicated the decision was his and he overruled Kraft by directing the reentry with the retrorocket attached to retain the heat shield in position. The lack of a timely decision prevented instructions to the Hawaii and California CapComs and as a result, Astronaut John Glenn was not made aware of the heat shield issue and reentry configuration tradeoffs until contacted by the Texas site during reentry.

Glenn's mission and the three that followed: Mercury 7, 8 and the day-long Mercury 9 mission in 1963 satisfied the relatively simple Mercury Program objectives. (A) Place an American in earth orbital flight. (B) Investigate man's performance capabilities and his ability to function in the environment of space. (C) Recover the man and the spacecraft safely.

At Holloman AFB, I worked in flight test and led a team supporting the B-52 testing from King 1. The comparison between my King 1 test team and the Mercury Control team operations provided a wakeup call, that the mission command and control performance of Kraft's team was totally unsatisfactory. The only thing that held us together was the stalwart performance of Chris Kraft, and even then luck was on our side. The Mercury program was considered a success. We placed an American in orbit and returned him safely to earth, but from my standpoint the Mercury Program was a "boot camp" for space operations.

The most critical need for Gemini was the formation of a dedicated and highly skilled operations organization. The second was the development of accurate technical knowledge of the space systems, operating procedures, mission rules, and a training program that prepared controllers for rapid response to flight problems. The final action formally established the Flight Director as the sole individual responsible for crew safety and mission success during mission conduct.

9

THE TEAM

It always seems it is impossible until it's done.
—*Nelson Mandela*

When I grew up, I was a Detroit Red Wings hockey fan and Gordie Howe was my hero, just like Wayne Gretzky was with the Edmonton Oilers in my later years. As a sport, hockey is about great teamwork, of passing, of blocking, and of deflecting shots and assisting on a goal. The teamwork is fast-moving, relationships between players intuitive, and goals are celebrated by all. A hockey team must have "IT!" and a modicum of talent to succeed in the playoffs.

In November 1961, NASA announced that the STG was renamed the Manned Spacecraft Center (MSC) and would be located in Houston, Texas. This decision was made at the midpoint of Mercury, prior to the Glenn Mercury 6 (MA-6) mission. In September 1962, when the MSC was relocating to Houston, we packed the family into our black, 1959 Plymouth station wagon, attached a U-Haul trailer, and drove to the Cape from Langley for launch. After Schirra's MA-8 mission, we continued driving west to our new home on Welk Street in South Houston. I had six months to get the family settled in a new home, organize our new office location, get back to the Cape, and prepare for the final Mercury Atlas 9 mission with Cooper.

MSC underwent continuous change in support of the Gemini and Apollo programs. Kraft, now Director of the Flight Operations Division, was building divisions for Mission Planning, Flight Support, Flight Control and Landing and Recovery. Hodge was the Flight Control Division Chief with branches for Flight Control, Facilities, and the Mission Control Center. Twelve temporary locations were established in South Houston. Shopping centers, banks, TV studios, warehouses, and hangars at nearby Ellington Air Force Base became offices. Hodge's division was located on the Gulf Freeway, between Houston and Galveston, in a large vacant warehouse, with my Flight Control Operations Branch (FCOB) located in the concrete showroom display area, cluttered with government gray desks and chairs. The constant ring of the single phone quickly became a source of irritation for all in the room until we installed a switch to silence the phone. Kraft's offices were a short two-minute walk to the nearby Houston Petroleum Center.

The McDonnell Gemini spacecraft briefings introduced us to the new space technologies, digital command, spacecraft computers, cryogenics, fuel cells, bi-propellant thrusters, and inertial navigation systems. The Gemini goals of rendezvous, docking and EVA provided the key objectives essential for missions to the moon. The Gemini two-man crew used ejection seats to escape the highly explosive and toxic hypergolic propellants of the Martin Titan II rocket for a launchpad and early flight abort. Transition to the Gemini spacecraft reminded me of my transition from training aircraft to the F-100. Space had entered a new world of performance. In August 1964, I was assigned as a Gemini Flight Director along with John Hodge and Glynn Lunney. Lunney, the Bermuda Flight Dynamics Officer, was now forming the Gemini trajectory team for Mission Control. The Gemini Mission Control team had three major functional elements. The Trajectory team, known as the "Trench" consisted of the Flight Dynamics Officer (FIDO), Retrofire Controller (RETRO) and Guidance (Guidance) Officer plus two Booster systems Engineers (Booster) that supported the launch phase. The spacecraft systems team consisted of two engineers, supporting Gemini Guidance, Navigation and Control (GNC) and Electrical, Environmental, and Communications (EECOM). A single position supported the Agena rendezvous Target Vehicle (Agena). The third group is what I called the command and staff and includes the Flight Surgeon, Assistant Flight Director, Procedures, Network, CapCom and Flight Director. The console positions had two to four back-room controllers in support staff rooms (SSR) that provided the control room team specialty support.

The period from May 1963 to 1965 was solid work; there were no weekends. As I walked the warehouse floor, I could see and feel the space team forming. The whole room was filled with the murmur of the working groups assimilating knowledge of the new space systems technology. I celebrated when I was assigned a branch secretary, Maureen Bowen. Her office was at the end of the second-floor hallway in a room with two oval windows that looked like a movie house ticket entrance. Maureen reigned over one of the rooms and in the next was the Ditto machine (an early type of copier). Maureen did not mince words and in just a few days she demanded, "Either move me or move the Ditto machine." Maureen won, the machine was moved, and this began her reputation as "Mo." Maureen was a joy, talented and an expert at working in the formation of new organizations. She served as an acolyte fostering "IT!" Due to her prior work experience, with the HUD predecessor in Washington D.C., she was superb at organizing and within a month, I reassigned her to work with Brooks establishing the Gemini Systems Branch.

Bowen was one of the pioneering members in Mission Operations. She laid the groundwork for seven branches, becoming a "den mother" for the startup of new operations elements. Maureen received unwanted attention from the FBI and NASA Inspector General for her management of the design and finances of the "Apollo 11 Mug project." Her work violated NASA directives on the use of the Astronaut Badge and the location of division cash funds. The Apollo 11 crew member, Mike Collins interceded on her behalf at Headquarters, providing permission to use their Badge and the crew signatures on the mug. Maureen was the "go-to" person for most organization activities and in 1994 was assigned as the Directorate Administrative Officer.

Don Bray, another new hire, reminded me of a military supply sergeant. He was a skilled technician and scrounger. I assigned him to build a high-fidelity Gemini cockpit trainer to provide the "hands-on" feel for the instructors and flight controllers. Working with the McDonnell engineers, he obtained Gemini factory instrument panel drawings. The trainer became an after-hours job as additional engineers joined Bray's project. McDonnell provided reject spacecraft switches from the plant and the concrete office floor was soon littered with power cables. I believe we had a working cockpit trainer before the astronauts had one, so we began training controllers on the crew tasks. Bray's trainer continued to increase in fidelity and technical sophistication with no cost to the government. Within a few months, Bray began building sophisticated cockpit trainers for the Apollo Command and Service Module (CSM) and Lunar Module (LM). The original LM trainer stands in the visitor center today.

My replacement secretary for Maureen was Sue Erwin, a cowgirl barrel rider who competed in the Houston Rodeo. She was short and lean with calloused hands from riding and was intolerant of salty language. Sue was great in the office but she needed help with her son. He sometimes came to the office for an entire day. I talked to him like a father, exploring issues around schooling, discipline, and about getting along with his friends. Mostly I just listened—that is all that really mattered, he just wanted to be heard.

At times, as branch chief I felt like a Catholic priest listening to confession, taking in the troubles of others. I did not mind. The capacity for empathy is critical to strong leadership. A leader should never lose sight of the fact that an organization is not a "thing," and it doesn't function like a machine. Organizations are made up of organic people, each one of them different than the next, each beset by his or her own constellation of joys and looking to find their place and time to grow.

To run an organization, you need to see the minutiae as well as the whole, and you need to project understanding and compassion. I can't stress that enough. "Toughness" is not insensitivity and "Competence" is not about being robotic. Both principles call upon a leader's humanity.

As the chief of the FCOB and acting chief for the Gemini and Apollo Systems branches, I was building the Operations Team for Gemini and the pending Apollo lunar assault. To avoid recruiting overlap with other NASA centers, MSC was assigned the Central United States Great Plains as their recruiting zone. Texas and Oklahoma, known for cattle ranching, have mountain and desert terrain that eases into the stunning landscapes of the Great Plains. Some areas are extremely flat, while other contain tree-covered mountains, low hills, and streams. I had flown over the land as a pilot and its vastness formed images of the early frontier days. As I read resumes of new applicants, the words resilience, grit, strength, and fortitude came to mind.

I had recruited a suite of potential supervisors. Mel Brooks, a Korean Infantry veteran with Lockheed Agena experience; Jim Hannigan, an Air Force flight test engineer; and Joe Roach, an Air Force Captain in the Alaskan Air Command. All had seen Mercury program service to provide introduction to the world of space. Arnie Aldrich, continuing his work developing the Gemini spacecraft system manual, was assigned to lead the Gemini systems Section.

James Bates, a 1962 graduate from Southwestern State College in Weatherford, Oklahoma, responded to a newspaper ad. After being hired by NASA, he began conveying additional student resumes to NASA. The resumes came from children whose grandparents were the German and Russian settlers of the Oklahoma Territory. Oklahoma became a prodigious provider of astronauts, Flight Directors, controllers, and instructors in the early years.

On one application in 1964, the words "physics, math, rancher, and preacher" captured my attention. John Aaron was born in Quail City in the Texas Panhandle near the Oklahoma border. While in the military, I had traveled that country on Route 66 and found it was a tough, hard region; part of the Great Plains Dust Bowl. I liked the sound of this man and left the Form 57 with personnel before I departed to the Cape to observe the Gemini 1 Titan launch.

With the transfer to Houston, many of the CapComs who supported Mercury decided to remain at Langley. With "Dutch" now developing the Gemini Mission Control Operations handbook and closing out the remaining Mercury work, I spent significant time reading applications and recruiting. The greatest need was developing NASA leadership at the remote sites. I had hired Dan Hunter as a civil servant from Philco and needed him to augment my NASA leadership strength. Hunter, while strongly opinionated and with an insufferable ego, was a "take-charge" leader who did not wait for direction. In Korea, Lt. Glen Magathan had great flying skills but was a solo performer who never worked to develop his wingmen, leaving both vulnerable. I saw the same attitude in Hunter and matched him with Aldrich for the first two Mercury missions at the Guaymas, Mexico, site for on-the-job training to see if he would develop the team skills to perform as a CapCom team leader.

There were plenty of applications from recent college graduates but I was in dire need of leadership. My flying experience provided a high degree of respect for the Ground Controlled Approach (GCA) operators who talked you down in bad weather. A GCA team is housed in an operations trailer supported by a power and communications trailer located alongside the runway. The aircraft position in relation to the runway is provided by a search (long-range) and precision (short-range) radar. The facility has operator scopes and communications systems and generally can handle two simultaneous contacts, although once in my case they brought a flight of four down. When they provide the runway weather, they know it because their trailers are in the middle of it. GCA controllers' confident, steady voices provided course and glidepath during a weather jet penetration. With lightning, heavy turbulence, and with the aircraft automatic direction finder (ADF) needles bouncing between lightning strikes and the runway beacon, GCA quickly became your best friend. On two occasions after my landing, I left a bottle of the best whiskey available at the Officers Club for the team.

When I saw an application from Ed Fendell and the word "GCA," I did not even turn the page to see if he was an engineer. Fendell's application indicated he was a Pan-American GCA operator at Gander, Newfoundland, and a Korean veteran. That was all I needed to know. In dealing with people, my first move was always to appraise who they are, the source of their character, and what experience they would bring to the job. Then I would look at their education and grade point average. What I had learned from Mercury was that I would need "Tough and Competent" resilient workers, not students, for my Gemini team.

Ed became my "go-to" guy on many difficult and critical assignments in the coming years. He had his first test working in "secret" with me and a small engineering team developing the EVA capability and the supporting data for Gemini 4, the second Gemini manned mission. The plan was imaginatively labeled "Plan X." He was deployed to the Carnarvon site to give the astronauts the "go to depress" the cockpit to begin the EVA sequence. Later, when assigned to the Assistant Flight Director (AFD) for Gemini 9, he considered it a "non-job." He was loud, raucous, and pestered the flight surgeon on the console in the row to his front, occasionally triggering my call "Ed, knock it off!" I valued his fire and considered him an organizational asset, but the other Flight Directors considered him a distraction. He had the ability to communicate and work under pressure and that was a skill in short supply in my initial NASA cadre.

Tough and Competent Gemini controllers Ed Fendell, front and Charlie Harlan

Beginning Gemini, Ed faced a challenge that would provide the foundation for his career. A book of electrical schematics was dropped on his desk by Dutch von Ehrenfried, his supervisor. Fendell looked up and said, "I don't know how to read electrical drawings."

Dutch very curtly said, "Go to the bookstore and buy a book on basic electronics and begin color coding the various schematic symbols and functional components. That is how you learn in our business." With a two-year degree in marketing from Becker Junior College, Fendell made up his mind to learn by doing.

There were no Gemini manuals to study at night, no instructors to coach you through the first flight, however the open warehouse floor provided an ideal location for Gemini "bull sessions" to start building the unity, teamwork, and the personal relationships necessary to establish trust and build the team. All too soon, the Gemini Program was upon us, and it was hard to believe that the Apollo missions were just three years further on down the schedule. While I was building my three branches, Hodge had established the Simulation and Mission Control Requirements branches to develop and test the controllers' skills and implement the control center console functional display and control capabilities. The Gemini and Apollo schedule overlap became very apparent when I had to compact work areas on the warehouse floor to accommodate personnel from North American, Grumman, Agena engineers, and a Saturn launch team from the Marshall Space Flight Center. In a literal sense, we became "close." The building was a warehouse with limited sanitary facilities and was designed for a much smaller workforce.

1965

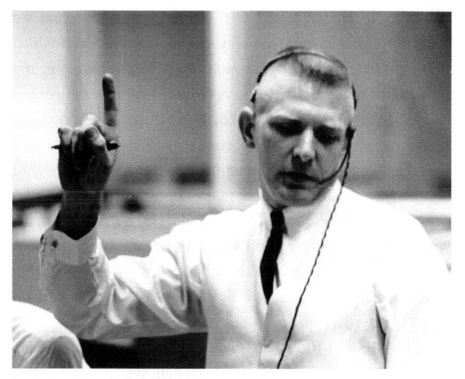

Kranz in one of his signature vests, all handsewn by his wife

Leaders often use a "signature" to form an organization in their image and dominate the field of play of an endeavor, one that establishes a character for the units they command. General Patton had the twin Ivory-handled .357 Magnum Smith and Wesson pistols; General McArthur had his Missouri Meerschaum corncob pipe; Carrol wore colorful suspenders and bow tie; and Saylor wore an Aussie Overland hat. A fighter pilot's scarf serves as a signature but also has the important purpose to protect the neck when one is searching the skies for aircraft. My wife made scarves for my flight in Korea. When Kraft, Hodge, and I were named Flight Directors we decided to identify our teams by colors: red, blue, and white. Prior to Gemini 4, my first mission as Flight Director, my wife Marta suggested wearing a "white" vest to establish my team's identity. The value of "identity" is that it purposefully establishes your brand. When I started my first on-console shift with the "white" vest, von Ehrenfried, sitting nearby, thought I was crazy. He had the camera in the control room zoom in on my console. The picture with me standing by the console in the white vest was displayed on all control room TVs, the press pool, and at headquarters. The "white team" legend and its identity part of the chemistries of a high-risk operations team with "IT!" was born.

The Mission Rules for Gemini now addressed a more dynamic range of mission phase dependent systems, trajectory, and data issues. Systems operating limits and reentry heating constraints were documented and telemetry measurement requirements for critical decisions were defined for each of the spacecraft and ground systems. Weather requirements were defined for launch and mission landing areas, and for the first time we really had some "What do I do if . . . Guidelines." The rules for the Gemini 4 mission consisted of about 120 pages.

Satellite communications now provided real-time direct data transmitted from Bermuda and California. Controller teams now only manned stations at the Canary Islands,

Australia, Hawaii, Mexico, and two ships thus providing more effective real-time support.

The continuation of "learning by doing" developed skilled systems engineers and the MCC computer systems provided trend data displays improving in-depth analysis capability. With the advent of dynamic real-time mission simulation, the controller teams were well on the way to becoming a highly effective asset to assure crew safety and mission success. It was time for my control team to fly!

10

CHALLENGE

Surround yourself with good people; surround yourself with positivity and people who are going to challenge you to make you better.

—*Ali Krieger*

The Houston Mission Control Center (MCC) and Simulation facilities were in concurrent development during preparation for the initial Gemini missions. As a result, during mission periods there was a shifting of personnel from facilities development to support training the operations team. I believe this was very beneficial in the knowledge transfer of the hands-on controller task to the development teams. As a result, with the completion of the simulator development, the personnel readily transferred to operations support. Complicating the work was the parallel development of the Apollo requirements for Mission Control, training facilities, and the training program development.

The sixteen-month gap between Mercury and Gemini was a tough period for all, especially the new recruits as they had to adapt to the long, pressure-filled hours with the calendar relentlessly moving toward launch day. FCOB provided most of the control center and remote site personnel and unlike the Mercury experience, I had solid leadership and skilled personnel in place at the remote sites that was now combined with a control center capable of processing real-time telemetry.

The greatest asset came in our mission training through a mission simulator capable of creating the "real time" environment, and the Gemini cockpit trainer for replicating the crew task. The trainer was located on the warehouse floor where we worked each day. This simple plywood cockpit with cockpit panels, subpanels, switches, circuit breakers, and lights was a major stimulant, a perpetual reminder that brought the program close to all seated on the 150 by 80 -foot warehouse floor. Controllers would compete for time to climb in to exercise procedures and commit switches and controls to memory.

I often thought of the risks we faced in aircraft flight test, but in spaceflight the risks increased by an order of magnitude. One day when we reached the moon, we would explore and then reach further. Maybe it was the work on the boundary of risk, with death ever present on the horizon that created and sustained the team chemistry? I often experienced this with my flight in Korea and now on the edge of Gemini operations, I felt "IT!" once again.

"Learning by doing" worked well in developing the controllers' individual skills and now it was time to test our readiness. A scrimmage tests a football team's skills and now with the mission simulator, the controller team relationships within and between disciplines could be tested and smoothed out. With the simulated spacecraft moving five miles/second, it became possible to exercise decisions during tracking station passes. The simulations formalized the procedural dependency between team members,and astronauts, welding technical as well as personal relationships. Team handovers between shifts ironed out irregularities and provided the needed consistency for events spanning days. Simulations provided a depth to our work in building procedures and mission rules and clearly identified the vulnerability to error. Unlike prior teams I had developed and worked with, this team would be with me for life. Team chemistry was forged and became a reality on the Stahl Myers warehouse floor and during mission simulations.

With skilled leadership in place and controller teams who had developed the plans, written the procedures, established the rules, and tested their abilities in simulations, it was time to step forward for our first flight. Unmanned flights are tough, but this one was a simple lob shot of about twenty minutes. Our only action was to transmit commands to back up the spacecraft mission sequence programmer. I had been off console for quite a while and a desk does not fit me well, so I was glad to get back into action at the Cape. Our new Houston facilities would be utilized in a backup mode. If Mercury was our boot camp, Gemini was where we would prove we had The Right Stuff to meet the challenges on our march to the moon.

The scheduled operational readiness of Houston Mission Control would not meet the initial Gemini Program launch schedule. Teams deployed to Houston Gemini Control, at the Cape, and the worldwide network for an eight-day integrated test of the control center and remote sites in October 1964. Computers had been introduced at the tracking stations to provide the controllers an ability to transmit telemetry snapshots to Mission Control, reducing controller workload, and improving real-time assessments. The personnel deployed were Mercury-experienced, with the prime duty to train the new personnel, test the new systems, and develop site operations procedures and training materials.

The testing went well and on return, the debriefing filled many notebooks of needed actions. It was all positive and we celebrated with the controllers, families, and provided the kids the opportunity to fly the Gemini procedures trainer. Unlike Project Mercury, the controllers were full-time and committed to operations. It was their job, their life, and after the missions' celebrations we were like one large family at a picnic.

January 1965

The Gemini Titan-2 (GT-2) was an unmanned ballistic test of the Titan and Gemini capsule, and GT-3 was a manned, three-orbit mission. Kraft with a small team would control both missions from the Cape. Hodge, with another team, would "flight follow" from Houston to assess readiness of our new facilities. Since Kraft and I would be at the Cape for an extended period, we leased an apartment in South Cocoa Beach. This period fortified the relationship we developed in Mercury. Our discussions ranged widely from program policies, NASA leadership, astronaut personalities, and the character qualities necessary to manage risk. Chris was Episcopalian and often spoke of the importance of prayer in a person's well-being.

Up close, I saw the full range of Kraft's personality as a leader, teacher, comic, and student. He exhibited every emotion in the process—inquisitiveness, anger, worry, and joy. There are a hundred words to describe him but in all cases he was thoughtful. His sayings provided a glossary of his personality, leadership, and character. As he talked, the timbre of Chris's voice changed to match his words and thinking. We worked in parallel: Kraft spending most of his time with the Headquarters program management addressing the downstream missions, while I attended to the Cape operations and mission team preparation for the two coming launches.

I am a "note taker," especially those words and phrases that involve leadership. The weeks sharing an apartment with Kraft provided my first in-depth grasp of the depth of Chris's Christian faith and how it influenced his thoughts and actions as a leader and my own continuing development. His words, "Do not be a stranger to yourself," and, "If you live by what you stand for, you will see it reflected in your team," were carried throughout my career. Each evening, Kraft and

I would debrief in the apartment or over dinner at Ramon's Restaurant, famous for its seafood and Caesar salad.

In a time before the internet, I received daily calls from my Houston team on mission preparation and deployment status, and the growing list of the many open items for the first manned launch. Kraft had never liked the ejection seats and as the troubles mushroomed in thruster and seat testing, fuel cell qualification, and target vehicle schedules, he became obsessed with assuring that his control team was "on top of the problems" and was "capable of detecting any problems in flight." I had been marching my controllers in this direction for a long time and with Kraft's direction, we had the blessing to do anything needed to get on top of our job. Aldrich burrowed deeply into the design and testing at McDonnell, Weber Aircraft (the seat subcontractor), Rocketdyne, and General Electric. Aldrich had a "gut instinct" as an engineer in finding and addressing problems.

The GT-2 mission was the first "shakedown" cruise for the Gemini launch team and the new Mission Control positions for the booster and guidance operating positions. We deployed to the Cape and both tracking ships nine days prior to launch with the full control center team, simulation team, and teams for each ship. Six days before GT-2 launch, Kraft developed a cough, headache, and dizziness to the point he was physically unable to leave the apartment. I supported both days of launch simulations, launchpad tests, prelaunch briefings, readiness review, and launch pre-count.

January 19, 1965

Kraft's recovery for launch day was one of pure willpower. He was impeccably dressed for launch day in a crisply starched shirt and tie but except for milk, he avoided food from the "roach coach" outside Cape Mission Control. To our surprise, we were not pre-briefed that the media would be allowed in the control room for launch. The countdown proceeded smoothly and seconds before T-Zero and liftoff, the control room turned brilliant white as the media cameras with their lights were turned on. This was followed by an instantaneous control room blackout except for the battery-powered console intercom panels as the Titan lifted off. In absolute darkness, the Mission Control technicians stumbled around and I tried to read my stopwatch by the glow from the communications panel to call out the backup mission sequence commands we had planned to send. We knew the Titan had lifted off from the ship and Houston Control callouts, listening in the dark as the reports came in from Lewis and Fendell on the ships. Hodge's crisp English voice seemed to enjoy reporting the flight status to Kraft from Houston Mission Control.

The debriefing was short and when Hodge told Kraft he should carry a flashlight with him to the Cape for GT-3, Chris did not respond but took his headset off and slammed it down on the console. The GT-2 "lights out" event led to the Flight Directors traditional "Battle Short" call to physically block the building power breakers, lock control room doors, and select recorders to flight speed. From that time on, all media cameras had to provide their own battery power.

Kraft and I returned to our apartment at the Cape for eighteen days to prepare for the first manned Gemini, GT-3. I found myself doing "double duty" as assistant to Kraft and as a Flight Director in training. This provided time for an in-depth shakedown of the control center, network systems, and teams to build integrated team confidence.

On March 12, 1965, two weeks prior to the first manned Gemini launch, the flight controller training program I had established was formally recognized in an MSC published news release. *"There is a school at the NASA Manned Spacecraft Center here, from which no one ever graduates and there is no summer vacation."* The note closed with, *"There you hear regional inflections from crisp British to Southern drawls and standard American where judgment and knowledge is gained by living a career in a continuous process of learning as a flight controller."*

The training program was one I had pursued since first joining NASA. For the first time, I felt comfortable with the total Gemini mission staffing. I knew and had worked with every member of the teams we deployed and with the closure of half of the Mercury stations; I no longer sweated out the "hired hands" performance. Flight Control now had teams with the professional capabilities and staffing for manned spaceflight operations.

Kraft had also established the Flight Director as the mission authority for all aspects of crew safety and mission success. Aldrich had developed the space systems technical handbooks, and we had meaningful mission rules supporting critical decisions. In the short period since Mercury, we had assembled a highly professional operations team.

Gemini continued the Project Mercury tradition that astronauts would staff the CapCom positions in Bermuda, Australia, Hawaii, and California to provide communications with the crew during launch, critical events, and deorbit preparation. Astronaut Conrad, a member of the second astronaut class, was deployed to the Carnarvon, Australia, site for Gemini 3. He arrived in time for the final network test on launch minus four (L-4) days. I had deployed Dan Hunter, a Mercury-experienced CapCom and FCOB section chief to the site as CapCom and he had supported the site checkouts and network simulations. CapComs have two principal duties: they must interface with the site manager for technical support, and communicate with the astronaut crew. Site passes range from six minutes down to possibly sixty seconds or less. The process of acquiring the spacecraft, processing the telemetry, issuing commands, and communicating requires effective teamwork within the site.

When the L-3 mission readiness review was completed, the launch countdown was initiated. Kraft and I celebrated with a dinner at Ramon's, then retired for the night. About midnight, a pounding on the door awakened us. When I emerged from my room, Kraft and Slayton were in a loud and angry argument over a conflict between Hunter and Astronaut Conrad over their respective roles at the Carnarvon station. The following morning, I called Hunter on the conference network and he indicated Conrad had told the site personnel that he was in charge. I briefed Kraft on the site problem and then wrote a message to the site defining the individual site responsibilities. My message assigned Hunter the "site" operations responsibility and Conrad the "crew communications" role, thus straddling the issue between the two men at the site.

The following day, L-2, the Network voice was called up for the last-minute briefings. Hunter, with Conrad at his side, opened with comments on my message stating, "This message does not clarify anything." After a few more choice words he finished, "When I get home, I am going to frame it and hang it in my crapper!" An angry Kraft then closed the briefing with the curt words, "You've got your orders, young man." I believed Hunter would relent and perform his duty professionally to manage the site and let Conrad handle crew communications.

The Mission Control team members, hearing the voice exchange with Hunter, picked sides . . . astronauts versus controllers. Hours later, this split carried over to the "Beach House" where we often partied with the crew after the final test. When words continued and challenges were issued in a loud face-to-face exchange between John Llewellyn and astronaut Alan Shepard, I corralled Llewellyn and departed with the controllers to the hotel. Living in an environment that combined the temperament of a football training camp and the confinement of a submarine, with ego and pride all around as well as relentless pressure, was hazardous. Only the threat of discipline prevented an occasional bloody brawl from breaking out.

GT-3's three-orbit mission with Gus Grissom and John Young was highly successful. Pride and ego, however, resulted in unacceptable Carnarvon station performance, resulting in loss of telemetry and voice during a spacecraft pass. After the controllers return and the debriefing, I fired Dan Hunter for his leadership and judgment failure. Kraft arranged Hunter's transfer to Goddard and during the Apollo years, we worked with him in his assignment as the Madrid station manager. Hunter was my responsibility, and I was accountable for letting the issue fester and failing to resolve the issue with Kraft and Slayton before the liftoff.

The lesson that I carried for life is that "There is no such thing as a small personnel problem!"

A week after Gemini 3, I was called to Kraft's office. I was mentally expecting Kraft to ask my readiness for duty as a Flight Director when he said, "If we can get the equipment ready, we are going to try for an EVA (extravehicular activity) on Gemini 4." I was stunned as he continued, "I want you to work with engineering, prepare a flight plan, and mission rules. The EVA requires White House approval, and we want to keep it from the press. You are going to have to keep your work secret!"

I began two shifts, Gemini 4 prep during the day and after JSC closure, the secret preparation for the EVA. The ground rules planned the EVA during the stateside pass so we could monitor telemetry and record video. The EVA was planned, equipment tested, and training done in secret during after-hours sessions at JSC. The EVA would be performed in daylight with depressurization and hatch opening over Australia. I selected Ed Fendell to work with me on the plan and assigned him as CapCom at Carnarvon for the critical Go/No Go to initiate the EVA sequence.

The Gemini 4 EVA was highly successful, establishing an American space record for EVA duration. The success was misleading, however, as we would find out on later missions. This was typical Kraft, to look for opportunities to step ahead, recover from schedule upsets, and demonstrate the operational prowess of his team. He was a gambler!

The Gemini Program schedule ramped up and sustained a launch schedule of about ten weeks between missions during 1965 and 1966. It was not uncommon to have controllers training and planning for a subsequent mission before the previous missions had been completed.

Gemini 5 was my second mission as a Flight Director and to a great extent my coming of age as "White" Flight. The mission with astronauts Conrad and Cooper would set a manned spaceflight duration record. Neil Armstrong and Buzz Aldrin were my CapComs and we knew not how fate would bring us together on future missions. As mission durations increased, new system technologies were tested, and rendezvous profiles were developed for approaches, rescue, and docking.

There were several controllers in my office when I returned after the final Gemini 5 simulation day. Standing upright near my desk was a resplendent American flag with a gold fringe, tassels, and eagle at the top. I knew the flags were only provided to Kraft and the top-level brass. Don Bray, standing nearby said, "There is no American flag in Mission Control during our missions, so I requisitioned one. You should keep it in the office." The flag was displayed during Gemini in Mission Control, remained in my office until retirement, and is now at Central Catholic High School in Toledo, Ohio, my alma mater.

Kraft launched the Gemini 5, with shift handover to my "White" Team planned for the beginning of the fourth orbit. Fuel cell technology was now providing power for the increased mission duration. The chemical reaction of hydrogen and oxygen with a catalyst produced electricity. Gemini 5 was our first experience with fuel cells and cryogenics in orbit. Early in the flight, a heater failure in the oxygen tank caused a pressure decrease from 815 psi to 70 psi. By four hours, Gemini was powered down and began drifting flight. After two orbits and with a stable oxygen tank pressure, it was decided to continue the mission. With shift change approaching I asked Kraft what he wanted to do. His reply was short, "You're the Flight Director, it is your shift, make up your own mind!" I had come of age.

My electrical, environmental, and communications engineer (EECOM) was John Aaron, born in 1942 in Quail, Texas: a small city that is hard to find on a map, with a 2000 census population of thirty-three. John's father was a farmer and cattle broker, and his mother a minister. They moved in 1957 to a ranch Northwest of Reed, Oklahoma, where his mother served as a circuit preacher. John had seven sisters who were all teachers and he graduated from Southwestern Oklahoma State College in 1964 with a double major in physics and math. He began work in the space program to get enough money to build up a herd of cattle while intending to return to ranching. Once he began the work in space he never left. The small college in Weatherford was an incredible source of controllers as I was building the Gemini mission teams. John was assigned to Aldrich's Gemini Section with two good mentors, Ted White and Larry Bell. After getting his feet wet on Gemini 2 and 3, he was assigned to my team for Gemini 5.

Prior to the shift change, Aaron had been involved in the briefings done at handover and quickly brought me up to speed on the analysis of the problems faced. John, after a brief discussion on the physics of cryogenics at ultra-low temperatures, told me that the heat leak into the tank would gradually increase the tank pressure and make more power available as the mission progressed.

John Aaron, Apollo EECOM (Electrical, Environmental, and Communications Engineer)

He sounded eager to test his belief and at the midpoint of the shift we began a gradual power up providing opportunities for the flight planners to begin inserting systems tests and experiments all while tracking tank temperature and pressure. John was a country boy, and when he leaned over the backside of my console, I could easily visualize him leaning over the wood rail of a corral, sizing up the work to be done, recommending the action to take. There would be many meetings with John beyond the day we began to power up Gemini, each one memorable and signifying "IT!"

The controller performance reminded me of aircraft flight test, progressively expanding boundaries and assessing the properties of the space environment. Crew workdays were expanded to keep the program moving to prove technology; demonstrate Apollo capabilities; and achieve the planning, training, and operations performance needed for the lunar challenge. Most of my personnel were doing double duty: flying Gemini and planning for Apollo in their spare time.

Kraft, Hodge, and I were assigned to the first Apollo mission. Kraft planned to terminate Gemini support after the first rendezvous, and Hodge and I would leave Gemini on the subsequent two missions, Gemini 8 and 9. All remaining program objectives required establishing proficiency in rendezvous, docking and EVA would be supported by Flight Directors Lunney and Chalesworth.

The Gemini rendezvous target was the Lockheed Agena, an Air Force upper stage rocket fitted with a Gemini Docking adapter, and systems for command control capability from the ground and crew. On October 25, 1965, the Gemini 6 Agena target spacecraft failed to achieve orbit. Overnight plans were initiated to accomplish the rendezvous objective by combining two Gemini missions. Gemini 7, a fourteen-day mission with Astronauts Borman and Lovell, was moved to the pad and launched on December 4. A rapid pad turnaround was executed, and then Gemini 6 rocket and spacecraft were moved to the pad.

Procedures, flight plans, and training materials were developed combining both spaceflights and executed in Houston Mission Control with the tracking stations modified to support the dual-spacecraft mission. The lessons from

Mercury resulted in the professional operations teams and the "can do" attitude to keep the Gemini Program moving forward at a critical time.

Gemini 6, with astronauts Stafford and Schirra, lifted off on December 15, 1965. The rendezvous with Gemini 7 demonstrated a critical mission capability for the lunar program. The remaining critical mission capabilities involved spacecraft docking and extravehicular (EVA) capabilities. A second critical Apollo objective was accomplished with the landing of Gemini 7 after fourteen days in orbit, longer than the requirements of a lunar landing mission.

Hodge was the lead Flight Director for Gemini 8, a three-day mission with multiple Agena dockings and a complex EVA. Our crew was Neil Armstrong and Dave Scott. To assign team and training resources to Apollo, Hodge and I decided to fly Gemini 8 on a two-shift basis. The March 16, 1966 launch of the Agena and the subsequent Gemini launch were flawless. Six and a half hours after liftoff, Armstrong reported to the CapCom on the Rose Knot Victor (RKV) ship stationed off the West Coast of Africa, "We're docked, very smooth, no noticeable oscillations." Communications from the RKV ship were excellent and we checked off a critical program objective.

Unknown to the crew, a Gemini thruster began malfunctioning, initiating a slow roll maneuver. Believing the problem was in the Agena, the crew disabled the Agena attitude control system. When that did not work, and the spacecraft continued rolling, they undocked from the Agena and thrusted away. With the undocking, the Gemini was a much lighter spacecraft and the spin increased to almost 300 degrees/second. The crew was on their own.

Gemini mission controllers Aldrich, Griffin (standing), Hutchinson, Loe, Brooks

At that time orbital coverage and support were limited to Hawaii, two ships, and voice-remote relay from Tananarive, Madagascar, in the Indian Ocean. The Coastal Sentry Quebec (CSQ) stationed east of Okinawa, Japan, reported, "They have undocked . . . are spinning at a high rate . . . they have killed power to the orbit thrusters and activated the reentry thrusters." The CapCom continued, "The spinning is slowing . . . they have used one of the reentry control systems." As the

Gemini left the RKV, the mission was a mess, and we were entering a period where we only had two ships for support.

Hodge turned to me and said, "Your team is trained for reentry . . . bring them home."

The handover was seamless and the work building a "Tough and Competent" team now paid off. Within minutes we had messages circling the globe, alerting the recovery forces. We would only speak to the crew on three scattered passes. Forty minutes after undocking, we had a plan in place. Retrofire brought the crew down in the West Pacific landing area, a rescue aircraft was overhead at splashdown, and three hours later the Destroyer U.S.S. Mason arrived to recover the crew and spacecraft. If we had failed, we would have flown for another thirteen hours, critically low on fuel until reaching a landing area. Gemini 8 was an ultimate test of "IT!"

A lack of vision in planning the mission, developing the procedures, and training the crew and controllers caused the Gemini 8 failure.

When spacecrafts are docked, there are interfaces between the spacecraft structurally, and virtually between all systems of both spacecraft. In the docked configuration, the systems procedures and related training must be treated as a common system serving both spacecraft. A second critical error was that we did not trust the Agena. If you do not trust a space system, solve the problems to restore trust in the system. It was a simple lesson, one that I learned over and over again through my entire career.

When your gut says you don't trust something, listen to it, and then go after the problem until you have fixed it. Because if you don't, you will regret it.

Gemini 9 ATDA, the Angry Alligator

Gemini 9 was an equally difficult mission. An Atlas failure resulted in a loss of the Agena to reach orbit. The Agena Target Docking Adapter (ATDA), a simple backup target system was launched and achieved orbit. The two-piece clamshell nose cone failed to fully open, preventing the docking of the Gemini. I was the lead Flight Director and immediately assessed EVA options to release the nose cone. I was familiar with the design of the nose cone system and decided the risks of an untethered EVA and the potential for jet thruster impingement on the crewman was an unacceptable safety risk. I closed my shift with the clear decision that we would not attempt an EVA to release the nose cone.

Later that afternoon, I was called to a meeting of NASA top management to assess releasing the ATDA nose cone during an EVA. I provided an extensive briefing on the nose cone mechanism, the energy released if the strap was severed, and the safety issues related to an untethered EVA during station keeping. Charles Mathews, the Program Manager, overruled my decision and we were directed to plan and execute the EVA. We worked throughout the night to design the EVA procedures and address the safety issues as best we could. The following morning after the CapCom briefed the EVA plan to release the shroud, the crew indicated that they were tired and they did not want to waste propellant.

The following day, a highly ambitious EVA objective with an Air Force maneuvering unit located in the Gemini adapter was attempted. Fighting the suit stiffness, oxygen umbilical dynamics, lacking lighting, handholds and foot restraints, Cernan's sweating covered the helmet faceplate severely reducing visibility. Blinded and fatigued with his heart racing, Stafford directed Cernan to abort the EVA.

After the mission I conducted a personal debriefing. In retrospect, I believe my briefing to NASA management on the risks of an EVA to release the nose cone shroud was technically complete and correct. I believe my issue obtaining support during the briefing was one of reputation.

I was a virtual unknown except to Kraft. Those present at the meeting were all top level and had limited knowledge of my prior work and capabilities beyond the fact I had worked during Mercury and had flown a few missions in Gemini.

Reputation is an essential supporting element for top-level leadership. Reputation is the quality of character and judgment, passed by the written word, or word-of-mouth about a person's characteristics or abilities.

Better recognizing the "reputation" need, I exposed my key personnel at top-level meetings, reviews, press sessions, and in one-on-one management meetings to establish their reputation. While to the benefit of my people and NASA in general, this in some ways worked against my organization because "my best" were frequently used to fill critical needs within program offices at JSC and Headquarters. Still, it was the right thing to do.

The second critical lesson was to better appreciate the respective Mission Control and crew mission capabilities to address problems. Stafford's abort of Cernan's EVA was determined by a direct and continuous observation of all elements relating to Cernan's problem in space. Stafford could continuously hear the heavy breathing during Cernan's communications and recognize the need for urgent action to terminate the EVA. In Mission Control, I was a listener while Stafford was an observer. Preparing for future missions, I increased my time in the cockpit trainer, observed EVA training sessions, and during shuttle did a suited dexterity training session to better visualize the crew environment and tasks during an EVA.

The Gemini program prepared us well for our future work in space. We had developed organizational leadership and strengthened our abilities in Mission Control with learning by doing. The mission rules sessions became an incredible source for training and visualization of the "what-if's" and "What are we going to do about it?" I spent hundreds of hours late at night color coding my systems schematics and developing simple study guides so I better understood what the trajectory team was telling me.

With the dramatic increase in fidelity of the simulation systems, the team integration and performance took a dramatic step forward in all aspects of our work. The integrated controller and crew exercises became so realistic that at the end of each day's training you could feel the "IT!" When readiness is achieved, the desire for action springs not from thought, but from a readiness for responsibility.

THE MOON AND BEYOND

11

IT!

It's the team! Teamwork is what the Green Bay Packers were all about. They didn't do it for the individual glory. They did it because they loved one another.

—Vincent Lombardi

During the Gemini Program years of 1965 and 1966 was when we formed "IT!" President Kennedy's lunar goal now focused us on the urgency of our every action. While we were brutalized by the Apollo 1 fire, our momentum and our belief carried us forward. We were now a team, not in search of glory or personal success but propelled by the power of love. Our strengths had come from many sources, most of them small steps taken both individually and as an organization. There was no single event; it was gradual years of learning, testing, growing, and experimenting. We built the skills and relationships to achieve excellence in our chosen profession, spaceflight operations. Then came the day when we knew we were ready. We believed in ourselves, and we marched into the lunar arena.

Behind our family's house in Toledo was the Saint Agnes schoolyard. On long summer evenings, I watched the kids play street basketball. I was intrigued by my neighbor, Don Donoher, who spent as much time teaching the game of basketball as he did playing it. He was a gifted athlete and mentor, one who was able to teach without being critical and he taught all the right skills. I did not know the young boy I was watching would one day become the head coach of the University of Dayton's basketball team where he would compile a 437-275 record over twenty-five years. I did not know he would one day serve as an assistant to the legendary Indiana University basketball coach, Bobby Knight, at the 1984 Olympics where the American team won the gold medal. I certainly didn't know Don would be entered into the National Collegiate Basketball Hall of Fame in Kansas City, Missouri. The teams he coached at the University of Dayton were noted for their discipline, tenacity, and sound fundamentals—frequently besting schools with greater individual athletes.

In Chapter 8, I introduced you to Arnie Aldrich, the engineer I described as a "doer" and as an original member of STG back in 1961 who wrote the first flight controller handbook. Aldrich came through the ranks, developing the spacecraft systems engineers for Mercury, Gemini, and Apollo as a major contributor to the growth of the culture of mission operations. Aldrich possessed the same nurturing gifts as Donoher, that he was able to develop tenacious controllers into disciplined team players recognized in the

NASA engineering ranks for their outstanding fundamentals. My toughest job as a division chief preparing for Apollo was to tell Arnie that I was not recommending him to become a Flight Director because he was too valuable as a branch chief. I needed him to build the space system engineers for the future. He accepted my words and was in the controllers' ranks through Apollo, then he was subsequently assigned by Kraft as the Deputy Program Manager for Skylab. Later, after Aldrich had moved through the NASA leadership ranks, he became my boss as the Director of the Space Transportation System in 1985.

***

In 1990, at an event marking the 20th anniversary of Apollo 13, Aldrich became the first person to use the term "IT!" to describe the special chemistry of Mission Control. "Chemistry"—I can think of no better word. Aldrich knew of "IT!" because along with my other branch and office chiefs, we formed an organization of leaders who cared, inspired, coached, and taught by example. Our daily work environment and our culture established a contagious chemistry. It was imbedded in our work and expressed in all our relationships.

Aldrich used the words "We knew we had 'IT!' It was all built in as we had been working on IT! for years. 'IT!' is a concept that originated when we received the direction to put a man in space."

Our early years on Mercury were rough. We had been thrown together like a sandlot baseball team, with little consideration as to whether we could work well together. There was limited work continuity and we arrived at the consoles in Mercury Control as strangers to each other and to the mission. Chris Kraft, the Mercury Flight Director, was challenged when he had to form a team starting with just three controllers: Tecwyn Roberts, Carl Huss, and myself. Beyond that, he had to assemble the elements needed to support pad tests and launch. The embarrassing Redstone 1 "four-inch flight" was a sobering first step. We did not know what to do and we certainly didn't know how to do it. But whatever we were *going* to do, we knew that we needed a cohesive, goal-driven team to accomplish the task.

Kraft and I recognized there was a great need for training. During Mercury, we were daily left to improvise and even shift our various roles just to continue operations. The trajectory-related elements, interfaces with the launch team, and range safety were well-defined and covered by Roberts and Huss. Many of the remaining daily mission functions fell to me and that provided me with the responsibility to shape the team both technically and procedurally. My experience at Holloman provided me with the knowledge necessary to stay on the right track.

At NASA, I had sixteen Philco contractors experienced in the early satellite programs to "mentor" the young and inexperienced NASA engineers. The needed training was recognized early but was oriented almost exclusively to the astronauts. Our greatest need in the MCC was to establish a training "forcing function" to teach those who showed up for duty to address issues and work *as a team*. Because of the immediate Mercury training need, three members of the original STG simulation team, Harold Miller, Richard Hoover, and Dick Koos, plus a recent hire Melvin Brooks, a Korean War infantryman, traveled with us to the Cape. Their paper simulations consisted of problems written on pieces of paper delivered to controllers describing a problem and requiring an action, and when possible the written problems were supported by telemetry tapes driving console meters. Soon after the Glenn mission, Kraft selected Aldrich, Llewellyn, and Lunney from the remote sites to staff the Systems and Trajectory consoles. A rudimentary training program was developed for the remote site personnel and the first graduate, Charles Lewis, deployed to Zanzibar, Tanzania.

The debriefings and "bull sessions" after each day's activity stimulated a more personal form of growth as a team as did the "no-holds-barred" volleyball competitions after each day's training or pad test. The shared frustration with the constant rescheduling of tests and restarts, of rockets blowing up, and even the good times when we brought our families to the Cape for missions seeded the core growth of "IT!" The team chemistry was still on the distant horizon, but personal relationships were growing as was mutual respect.

Controllers and their families were separated for weeks at a time and for those assigned to ships in the oceans, months at a time. We celebrated the return to Houston: debriefed, traded gifts with family, told stories of journeys in primitive lands, of civil wars, and of life far from the things we take for granted. There was seldom a weekend off and while families were not forgotten, a different meaning and purpose held sway on our lives. Every day when we left work our mindset was the next day's work, open items and schedule, often at the expense of family communications at home. We were working like this because we believed in each other, the mission, and because we knew that without each other the mission would never be realized. Many of our families lived in South Houston on Welk and Regal Streets. Our area became known as "Flight Controller Alley" and after each day's work we continued the relations on the streets and front lawns as our children played on the grass or played kickball in the street.

Our office conditions brought further opportunities to merge as a team. The facilities were basic. We lived on a warehouse floor with sheets of butcher paper over the showroom windows to bring the temperature down to a mere ninety degrees. Secretarial support was limited, so we ran circles around the noisy and smelly Ditto machines ourselves, assembling the data package for the next mission even as we reeked of the machine fluid. There was no privacy, so personal conversations were taken outside, usually with a cigarette. The momentum of the work, the shouting of "Eureka!" as we made a discovery, and the sharing of knowledge was dynamic and all that mattered was "We were together on the front lines of space!"

Many of the original controllers were drafted from engineering elements at Langley and when they elected to remain in Virginia, their vacancies made room for new blood. The recruiting zone for MSC was the Midwest. Even though there were several large Midwestern universities, many of the job applications came from small state colleges. As we prepared for Gemini, I was staffing three branches. A new hire, James Bates, brought in twelve applications who were all physics graduates from Southwest Oklahoma State College. The applicants were sons of farmers, ranchers, and Native Americans born on reservations in the Dakotas. In those days, I needed workers, not students, and it paid off.

Organizational competition became a contributor to "IT!" Chili "cook-off" teams, Flight Operations Olympics, and division parties were family events engaged in without rank or rancor. The picnic dunking booth, sack races, and tag football competed with the inter-branch challenges. The astronauts and support organizations occasionally joined us and when I fractured my shoulder during an Olympics sack race, my flight surgeon accompanied me to the emergency room, though I still returned in time to hand out awards. The Kodokan Judo team formed by Dutch von Ehrenfried included several controllers, among them: Llewellyn, Moon, Bray, Lunney and others. In addition, we provided lessons in self-defense to the secretarial staff.

Traditions were established to formally retire Flight Director Team colors and to pass out American flags and cigars after each mission. A "Most Valuable Controller" was named after splashdown and given the honor to hang the mission plaque in the operations room. Three debriefings were held after the mission. The Crew-Controller debriefing concluded the formal mission reviews. Afterwards, a Crew-Controller party was held at the Hofbraugarten German Village restaurant in Dickinson, Texas. There, during hours' long monologues and a barrage of irreverent comments, the "Dumb Shit Awards" were given. You had to have a thick skin, or you did not belong.

Captain Refsmat, the "ideal" flight controller as drawn by Ed Pavelka

Bill Mauldin won a Pulitzer Prize during WWII for his cartoon characters, "Willie and Joe." They were foot soldiers whose everyday experiences became a favorite for the GIs in the field. Following in the footsteps of Mauldin, Ed Pavelka, a member of the trajectory team for Apollo and a gifted cartoonist himself, sketched out the "ideal" flight controller. The cartoon originally hung in his office, but soon office members outfitted the Captain with tools of the trade: a lapel insignia, combat ribbons, and a helmet with an attached antenna. The three-foot sketch was then taped on the side of a locker in the hallway and soon began to sprout graffiti. I would walk down the hall, read the graffiti, and gauge the health of my organization. Captain Refsmat and his arch enemy "Victor Vector" flew the missions with us and soon mission awards were sketched on his uniform or on lanyards over Refsmat's shoulders.

The Trajectory Team, commonly known as the "Trench," was located on the front row of the MCC. Behind the Trench were the consoles for the surgeon, CapCom, and four spacecraft engineers who "baited" the members of the Trench with statuettes descriptive of their job. An eagle proudly sat atop the EGIL (Electrical, General Instrumentation and Life support) console. During the long Gemini midnight shifts, there were communication gaps of well over an hour so we would order pizza or Chinese food, inviting the building security and support teams to join in.

This was much more than mere fun and games. The relationships we developed through such play were intimate, respectful, trusting, and functional. Initially, we were an all-male group and walked the gender boundaries with our secretarial and administration teams during the early period of cultural change, though that would change later on. The controller and astronaut post-mission parties at the Hofbraugarten were where we celebrated "IT!" We broke in our mission mugs and with no "civilians" present, we did a bit of carousing but we were never entirely off topic. We would critique individual and team performances, tell some jokes, review the glitches, and with our flight surgeons officiating, engage in arm wrestling with Jack Schmitt, the astronaut shill. We revived the losers, drank more beer, and continued debriefing. That is the special quality of "IT!" The fun was enmeshed with the work because when you are that close of a group, there is no separation between work and play.

Reputation, either written or by word of mouth, precedes an individual's entry into an organization. A reputation of trust allows time to work within a group, to build the shared values essential to teamwork, and to operate at high performance. Exposure builds reputation both within and outside the organization.

A leader must be respected as a person, possess solid judgment, be emotionally stable, and build and use his teams well. He must be technically competent, possess vision, and carry confidence. The process for the selection of leaders is a critical element in creating team chemistry.

Flight directors trained in Apollo to support Skylab and Shuttle Front. Rear L-R Milt Windler, Gerry Griffin, Pete Frank, Glynn Lunney. Front L-R Chuck Lewis, Don Puddy, Neil Hutchinson, Philo Shaffer, Gene Kranz

Over five decades, the Flight Directors, with only one exception, were selected from the ranks of the mission controllers. In preparation for Skylab and during the early Shuttle Program, I selected the Flight Directors in consultation with the existing Flight Directors and the simulation supervisors. The simulation supervisors possess a keen eye and develop a unique perspective relative to team building during the simulations and debriefings. A new generation trained in the late Apollo missions was selected to support Skylab, the Russian Soyuz Mission, and early shuttle missions.

Later, with my assignment as Deputy Director of Mission Operations, the Flight Director selection responsibility was assigned to the Flight Directors Office.

I was always under constant threat for not following the written civil service regulations, requiring separation of contractors and civil service personnel. To maximize cross training and exchange of knowledge, I saw them as one set of individuals performing the same or highly related work. The office floorspace allocation was based on the civil service headcount which meant, of course, that there was insufficient space for the contractors and this was the case even though there were regularly more contractors than civil service personnel. By lumping all personnel into one bucket, I knew that we had a better chance of achieving "IT!" if only because everyone would be working together on an equal basis. Nonetheless, I was forced to use the corridors for classrooms as well as space for the copy machines and storage cabinets. The systems engineers prepared the handbook schematics while standing in the hallways, using the walls as their drafting boards. The approach resulted in building safety issues but we never surrendered on the principle of "collocation," our term for putting similar skills in adjacent places. The supervisory responsibility was more complicated. The branch and section chiefs used the best and most experienced personnel as the leaders regardless of the badge they wore.

I required section and branch supervisors to support missions. To ease mission staffing and maintain supervisors as role models, the entire staff structure was required to support the testing, training, and operations. This required everyone to maintain proficiency as operators as well as organization leaders. During simulation and mission debriefings, all personnel—astronauts, Flight Directors, supervisors, and controllers—were required to participate and accept criticism for their performance from SimSup (Simulation Supervisor) and their peers. Ruthless honesty was a critical aspect in maximizing the benefit of the training, debriefings, and developing the all-important "IT!"

Team chemistry is hard to define but, like leadership, you know it when you see it. I have experienced team chemistry on several occasions, especially when leading the teams of Mission Control in efforts that demanded perfect time-critical team integration to address complex issues to achieve an objective. The timing was such that the talent and the fight we faced elevated the team's performance to the highest possible level. When Aldrich made the statement, "We had IT!" he was looking back at the decades preceding the Apollo 13 event where Mission Control members grew closer technically, emotionally, and socially in service to the team and where a conjunction of circumstances ignited the talent to perform as one.

Mission Control logo designed by Robert McCall

Fighter pilots rally around their squadron insignias and in my office, I had the patches of the 97th, 355th, and 69th Squadrons. With each, there are associations with missions and personnel. During the Apollo missions, space artist Robert McCall would often sit on the floor ledge just to the side of my console sketching the Mission Control team in action. I asked him to design an emblem to recognize the teams of Mission Control and the programs we supported. I provided him the following thoughts: the Latin phrase *Res Gesta Per Excellentiam*, "achievement through excellence" and an Earth with a satellite in orbit against a star as the background. Dutch von Ehrenfried proposed the Greek letter *Sigma* to represent "the team," and a rocket launched through the *Sigma* represented each controller's contribution. Mission Control was in bold, white letters against a red placard across the center, and four stars for discipline, morale, toughness, and competence on the border.

For five decades, McCall's basic emblem, with some modifications, has served as the anchor point of Mission Operations for all manned space programs. A corresponding document called "Foundations" defining the qualities needed for "excellence" in operations, was prepared in 1980 by Flight Director Pete Frank. It stressed the importance of discipline, competence, confidence, responsibility, toughness, and teamwork. After the loss of the *Columbia* Shuttle in 2003, the Foundations document was amended to incorporate "Vigilance," which was defined as "being always attentive to the dangers of spaceflight; never accepting success as a substitute for rigor in everything we do." Pete Frank's original Foundations document was short and to the point. So much so that it was rendered as a poster that, amended to include "Vigilance," can be found in offices, businesses, and schools around the world today. I include the "Foundations" in my response to students and individuals when addressing guidance questions on values. In 2010, the Kolbe Corporate Personal Growth Seminar borrowed from Foundations in its introduction when it included this line:

> The Foundations speak to a higher calling and sense of duty that we must accomplish the mission and derive value from our activities in space. They are as much about us as individuals as they are about the organization.

The beauty of the Foundations of Mission Operations is that these values were not dreamed up by a focus team on a management retreat. They grew organically out of our work as a team that had "IT!" They were values revealed through how successful human spaceflight was conducted for fifty years. When I think about Foundations, I think how each iteration of it—emblem, document, poster—was critical to establishing these common values so they would resonate in the workplace.

When I write about "IT!" and the flight control team, I am speaking of a single, complex living organization that is composed of 350 Apollo personalities, divided into five teams with interrelated planning, trajectory, medical, space systems, facilities, and communications duties but always tethered to the same core values. These people are always interacting with each other, often in split seconds, during a mission. They may be "many," but they act as "one."

In later years, for the Shuttle Program, my team grew to a NASA-contractor organization of over 5,600. I was responsible for the shuttle flight software, development, and operations of all mission facilities and conduct of all aspects of mission operations. I am convinced that no organization can undertake responsibilities of this magnitude and succeed unless it has "IT!," the critical asset that triggers a resolve during a crisis. The conjunction of the social, emotional, and psychological capabilities across the entire team raises team performance to the highest possible level.

While "IT!" may develop organically, that doesn't mean there is no need for attention and maintenance. For instance, because of consolidation and the enormous growth of our operations, communications became an issue. Under that kind of pressure, rumors and incorrect information can destroy the core of an organization, waste time, and kill momentum. For years, Mission Operations published a newspaper called the *MOD Focus*. Internal and external distribution of 2,600 copies assured all employees remained informed of the mission overall status and pending issues, and voiced opinions. The *Focus* was an internal publication and articles were solicited from all work areas. The name was appropriate as it was critical to keeping us all "focused."

Organizations, like humans, have a life cycle. They are born, grow, mature, and often begin to decline. I was fortunate to witness the birth and growth of Mission Operations as it reached a plateau of excellence. For many years, Mission Operations served as an example for others because of "IT!": team chemistry that inspired everyone to dig deeper, to find the last available resource for their team and teammates. Mission Operations succeeded in every role because it created an environment where the daily work was fun, where we were a family, and where we knew we were the best at what we did. In the post-Apollo period, several organizations were merged into MOD, each bringing its own culture and problems.

To successfully transition new organizations into MOD, I exchanged top-level leaders (normally deputy chiefs), to smooth out the transition, learn from the other organization's culture, and establish the technical and personal elements of "IT!"

Organizations are subjected to stress; it is the greatest threat to "IT!" and while I will analyze this issue in Chapter 20, I believe stress occurs because of internal or external forces, and usually from individuals lacking in character. Character counts, it is the "destiny component" in leadership. A true leader must exhibit a moral authority that wins personnel to their cause. For the leader, that authority must be coupled with a visible and uncompromising personal belief in the mission and the importance of the "team" in realizing that mission. They must believe they have been called to accomplish something greater than they would have been able to accomplish by themselves, that only the team can do it.

John W. Gardner, in the preface to his excellent book, *On Leadership*, states:

In order for an organization to function, leadership must be dispersed throughout all segments and at every level there must be individuals capable of taking leader-like actions to make their piece work. Men and women who are not afraid to send word back up the line that newly announced policies need amendment or reversal.[9]

Gardner then mentions the large numbers of people who are torn loose from previous values in what he calls the divergence of value systems. "Leaders are always seeking the common ground that will make concerted action possible. *It is impossible to exercise leadership if shared values have disintegrated.*"

The shared values that the Foundations document articulated for NASA—discipline, competence, confidence, responsibility, toughness, and teamwork—needed to permeate flight operations and serve as the means to achieve successful action. We had "IT!" and we felt we were ready to meet all assigned challenges.

12

TOUGH AND COMPETENT

In tough times, that's when you see true colors and personality.

—Didier Deschamps

I enjoy team building and speaking about the challenge of bringing a team effort together. I often use the word "mindset" in my discussions. To me, "mindset" is an individual's way of thinking. More than words, it consists of characteristics that establish a distinct and individual identity. Words are important in establishing mindset. People often ask if I was a fighter pilot and I respond, I was a *Sabre Pilot*, the key discriminator among fighter pilots who in our day flew the best, the fastest, the MiG Killer. Back in 1967, after the Apollo I disaster, when I stood before my controllers my goal was to drive the words "Tough and Competent" into my team. To override the shock and grief and replace it with the mindset that declared, "never again."

After the Apollo 1 fire on the launchpad, the NASA leadership—James Webb and Bob Seamans—appointed a highly skilled accident board. Congress had the foresight to stand down and let that team do what it did best without political interference. Our task was to ensure that the three men—White, Grissom, and Chaffee—had not died in vain and as the accident board looked for lessons to take away from the fire, there was simply no place for politics. In the end, many organizational and leadership changes were enacted.

I was assigned to lead the Flight Control Division. I had been working as Hodge's deputy, so the work was familiar but I needed to establish a structure to lead the division and the missions. I did not know when we would next fly but knowing the members of the accident board, my gut said it would be soon. My first step involved a tough decision. Recognizing the division and mission duties would be in conflict, I knew that I could not do both well while maintaining the standard of excellence. I elected to support the odd-numbered missions beginning with Apollo 5, thus establishing my Flight Director mission assignments for the program. In addition, the division needed structure for the complex tasks of accident investigation, recovery support, and the mission sequence changes.

I divided the seven branches into two groups—Operations-Training and Flight Systems—and established offices for the Flight Directors, finance, and administration. Fortunately, I had three excellent candidates to support the restructure: Jones Roach was assigned to Operations, Mel Brooks to lead the Systems and Experiments, and Glynn Lunney to lead the Flight Directors. Finally, as on prior work assignments, I needed tutoring on lunar trajectories so

I could hold my own when planning missions and as a Flight Director on console. Throughout Mercury and Gemini, Carl Huss, my roommate at the Cape, was my tutor. Now I asked Floyd Bennet, from the Mission Planning and Analysis Division, to train me on lunar-mission design.

Most of the Apollo systems technology was developed and proven during Gemini, and there were many parallels in planned utilization between the programs. One major difference, however, was that each of the two lunar spacecraft had a massive 64,000-word computer that would require maintaining by the Guidance Officer position in Mission Control to address the computer and related flight software. In Mission Control, the computer technology had raced rapidly forward to meet the challenges of spaceflight, and the IBM 709 evolved to the 7094, the first fully-transistorized computer. Now, in Apollo, we were using the IBM 360-75 which provided computing power supporting real-time trajectory processing and maneuver planning. In addition, it provided telemetry processing and display formatting for the systems controllers.

When George Low assumed command of the Apollo Program after the Apollo 1 fire, of his many concerns was the Massachusetts Institute of Technology's (MIT) performance in delivering the Apollo flight software. Low turned to Kraft with the directive to "make it happen." Kraft assigned the task to Bill Tindall.

Bill was born in New York in 1925 and grew up on the Massachusetts coast. After high school, he joined the Navy and served in the Pacific on destroyers. After graduation from Brown University in 1948, he worked for Goddard Spaceflight Center in software development and then transferred to STG at Langley. In 1960, within the Mission Analysis Branch, Bill Tindall initiated the development of space rendezvous capabilities. That changed when Bill was assigned to manage the Apollo flight software development. This assignment placed Bill in everyone's "sandbox" including my Flight Dynamics Branch.

Bill Tindall - Inventor of the Data Priorities

The allegiance to Tindall did not come easy for the trench. As the JSC technical community began development of the

Apollo software, the processes, requirements, complexity, and interconnectedness became obvious. Most organizations did not have the technical depth or were too parochial, so Low's program office invented the "data priorities" function and needing a charging, assertive, and integrating leader, put Tindall in charge.

The meetings often went on for days, the participants passionate and colorful, with plenty of opinions expressed. These men were not the kind to hold back. At times, little decorum was exercised because Bill was after answers. A couple of days after a meeting, a two or three-page memo would be distributed saying, "If I heard you right, here is what I think you said and here is what I think we should do?" Tindall's memos forced every program element to come to grips with their own personal devils and trade needed risk versus gain so the program could move on. Bill was the ace at his business in flight techniques, forcing us to step up to face tough decisions. In the words of astronaut Buzz Aldrin, "Bill had a brilliant way of analyzing things and the leadership that gathered diverse points of view with the utmost fairness."[10]

Tindall's data priorities addressed many arcane but important issues. When the moon's Sea of Tranquility was selected for the landing, Tindall alerted planners to the unusual conditions in that location. The LEM would start descent from about 60,000 feet above the mean (average) surface of the moon, but its *true* altitude above the landing zone would be 9,000 feet less. Moreover, it would be uphill because there was a one percent upward grade in the direction of the flight path. The data priorities meeting of March 7, 1969, closed out many open items for the descent some of which seemed trivial to outsiders. Throughout the descent trajectory, several Go/No Go decisions were made. Tindall suggested after landing, the terminology should be changed to something distinct like Stay/No Stay and then recognizing he sounded like a worry wart, added, "Just call me Aunt Emma."

Bill, already a proven leader, became a close friend with whom I shared a love for the ocean and for sailing. After missions, our families would often celebrate at his beach house on Texas's Bolivar Peninsula.

Tindall was the architect for the Apollo flight techniques. He possessed impressive observation powers coupled with an innate ability to find solutions for complex problems. I considered him so critical to the success of the Apollo Program that I invited him to sit on my left side at the Flight Directors console during the first lunar landing. In a letter to me on my retirement he stated, "I'll never forget the endless days of excruciating Apollo 11 descent sims and, most especially, your inviting me to plug into your console for the landing. This was a once-in-a-lifetime event for the both of us."

George Low's direction of the Apollo Program and the team he put in place initiated a swift recovery from the capsule fire. Frank Borman's assignment was key to the discipline needed to make the necessary systems, policy, and procedural changes. He was sick of the second guessing on the accident. As a Cold War "warrior," he wanted to beat the

Russians by conquering the high ground of space. The board was never able to conclusively identify the cause of the fire but provided the conditions leading to the disaster: a sealed cabin pressurized with 100% oxygen, extensive combustible matter in the cabin, vulnerable wiring and plumbing, and inadequate provisions for crew escape, rescue, and medical assistance.

The CSM Systems Branch led by Arnie Aldrich, was involved in every aspect of the spacecraft reviews. They participated in the configuration boards and incorporated all changes into the system schematics and flight procedures.

"Earthrise" as captured by Apollo 8

My gut feeling was proven right with the launch of the first Saturn V in November 1967, followed in 1968 by unmanned tests of the LM and Saturn V. These were followed by the Apollo 7 test of the CSM with Astronauts Wally Schirra, Don Eisele, and Walter Cunningham. LM development delays threatened the Apollo goal of a lunar landing before the end of 1969. To address this threat, Apollo Program Manager, George Low met with JSC and MSFC in August and proposed a plan for a December lunar mission. The December 21st Apollo 8 liftoff, with Astronauts Frank Borman, Bill Anders, and Jim Lovell provided mankind's first encounter of the moon on Christmas Eve 1968. I will always remember that Christmas Eve. It was then, in a televised transmission broadcast while they were orbiting the moon that the crew read the first ten verses of Genesis, concluding with Frank Borman's recitation. Borman then signed off by wishing everyone back home a "Merry Christmas."

Apollo 9, launched in March of 1969, was a ten-day manned earth-orbital mission to qualify the LM. After attaining orbit, the critical transposition, docking, and ejection was performed with the CSM docking to the LM and extracting it from the S-IVB booster adapter. The mission included numerous LM engine tests and an active LM rendezvous and docking. A planned EVA was shortened because the lunar module pilot, Rusty Schweickart, was ill. The Apollo 9 mission provided two critical lessons for the subsequent lunar missions. The communications management responsibility that was divided between the CSM and LM during Apollo 9 was poor, delaying decisions for mission events and requiring extensive troubleshooting. Flight experience indicated the need for centralizing all communications and data responsibility at a single location in Mission Control.

Ed Fendell was again my "go-to" guy to form the organization to get the job done, with readiness in ninety days. To organizations without "IT!" in place, this would have been impossible but not for ours, and as a result, not for Ed. The work involved selection and training of personnel, establishing control room operations capabilities, developing documentation, designing console displays and controls, and integrating his new team with other elements in Mission Control.

Apollo 9 was the first mission where the crew was in a spacecraft incapable of reentry. The CSM was the reentry vessel. A LM was docked with the CSM and when we separated the spacecraft, it had to be redocked to bring the crew back. The simulation supervisor quickly devised simulations that failed the LM spacecraft systems' ability to perform maneuvers and as a result could not, on its own, return to the CSM. Hence, the CSM must become the active maneuvering partner for rendezvous. This could take up to twenty-four hours, requiring the LM team to develop "power down" procedures to extend the LM lifetime and support the rescue. After the mission, the procedures were improved and formalized as part of a "lifeboat" option.

Additionally, the Apollo 9 docked LM descent engine test procedures executed during the mission became a part of the documented "contingency" options available to the control team. Apollo 10 provided a prototype for the entire landing sequence. This included a partial-powered descent to the lunar surface followed by LM separation and rendezvous with the CSM.

Mission Control had achieved the "IT!" factor and the incredible teamwork necessary for the journey to the lunar surface was on the horizon with Apollo 11. During missions, crises may arrive in many forms, some requiring rapid, critical, complex, and irrevocable action. In other cases, time exists to selectively take incremental steps, what I call "down-moding." This can also allow us to buy time to better define the problems and develop options. In both cases, the Mission Rules development discussions provided strong guidance to define the best path available. I often receive letters addressing the audacity and magnitude of our work at Mission Control. Each time I open one of these missives, it prompts memories of that day. Recently, addressing the National Air and Space Society while standing before the IMAX screen at the Smithsonian, I was once again reliving the details of the Apollo 11 landing. In preparation for my talk, I had integrated my console notes, crew debriefing and air-ground transcript with the mission rules that I used during the landing. The Mission Rules defined criteria for each of my teams' Go/No Go's. During the shift, my team of about nineteen controllers monitored both the CSM and LM but after the undocking GO, a second Flight Director covered Mike Collins in the CSM.

My landing team now consisted of six controllers who were a cross section of our nation.

My Retrofire Officer is Chuck Deiterich, originally from Bloomsburg, Pennsylvania, though he spent his teen and college years in Texas. Flight Dynamics is Jay Greene from Brooklyn, New York. Guidance is Steve Bales, hailing from Ottumwa, Iowa, where his mother worked in the town beauty parlor and his father was the school custodian. Steve later revealed that his inspiration to work in space travel derived from watching an episode of the *Wonderful World of Disney* when he was thirteen years old. The episode speculated that one day man would walk on the moon and it left a life-changing impression. "This wasn't the ordinary thing to do for a boy raised in a small, Iowa farming community," he said. Propulsion was Bob Carlton from Etowah County, Alabama, who quit school in the tenth grade and joined the Air Force. Communications Electrical, Life Support and Data is Don Puddy, from Oklahoma. My communicator is Astronaut Charlie Duke, who hailed from a Navy family, but was raised for the most part in South Carolina.

My crew is Buzz Aldrin from Montclair, New Jersey, and the West Point Class of 1951; Neil Armstrong of Wapakoneta, Ohio, an hour's drive from my boyhood Toledo, and a Purdue University alumnus; and finally, Michael Collins who was born in Rome, Italy, where his father was the military attaché. Collins also went to West Point, graduating in the Class of 1952 along with Ed White. I had worked with Aldrin and Armstrong before in Gemini and spent hours with them during training. There will be many split-second decisions and the fact that we know each other well will be crucial. Our trust must be absolute.

In an "All Hands" meeting two weeks prior to Apollo 8, the first lunar mission, I addressed the total division membership telling them I expected technical and operational perfection. To get there, they needed to be "tough enough" to admit and correct mistakes and "competent enough" to achieve success in all missions. I then concluded with their mission task,

You gentlemen will be making more unreviewed decisions than any similar organization in history. There is no flight test organization that has the latitude to make the decisions you make.

This statement would end up being prophetic for Apollo 11.

July 20, 1969

When our shift begins on Revolution 11 the crew is suiting up and then pressurizing the lunar module. The docked spacecraft are in an orbit sixty nautical miles above the moon's surface and my control team is looking over the crews' shoulders as they check out the systems.

A lunar revolution takes 118 minutes and revolution 12 passed quickly as the crew aligned the navigation systems and the ground uplinked maneuver data to the LM primary and abort computers for the planned maneuvers and abort targeting. The LM landing gear is deployed and near loss of signal (LOS), the undocking "Go," is passed to the crew. The LM had undocked and an inspection by Collins in the CSM indicated we had a "beautiful flying machine." Throughout Revolution 13, the final maneuver data was transmitted to the LM along with the backup voice data. We give the "Go" and then the CSM slowly separates from the LM.

I can feel and hear the room begin to tighten up. The adrenaline, no matter how you try to hide it, is really starting to pump. Prior to LOS, the LEM is given the "Go" for Descent Orbit Injection (DOI) which will be done behind the moon to lower the LM orbit to 50,000 feet for the powered descent. At LOS, I give the team a restroom and smoke break, but as they leave there is not the normal banter—no joking, only preoccupation.

Today is different, this is no simulation. Today we are about to land a man on the moon.

While Apollo 11 was behind the moon, the control room communications are rearranged for two teams to handle *Columbia* and Eagle in parallel. Sitting behind me are Chris Kraft, Dr. Gilruth, and George Low. Startled by a sharp rap against the glass I turn and glance through the window into the darkened viewing room behind me. The room behind the glass has seventy-four seats and they are occupied by what you might call the hierarchy of spaceflight: Administrators, Center Directors, MIT, astronauts, and many others. Indeed, on this day, the room is stacked like cordwood with people, some with their faces pressed against the twin-paned glass. Irrespective of their seniority, all sitting behind me are observers. The spectators earnestly try to decipher the terse, cryptic words on our communications loops. I turn back from them, set my cigarette firmly in the ashtray, take a deep breath and prepare to speak.

Handover to the Apollo 11 Landing team

My controller's average age is twenty-seven. Bales, my guidance officer, is celebrating his twenty-seventh birthday today. Sitting next to my console is Tindall—my friend and pioneer who assembled the software and hardware guidance, navigation, and procedural pieces together for this day. When my controllers return, I direct them to a private communications loop used only by the team. I feel driven to speak to them. I call it "our time."

Today is our day and the hopes and dreams of the free world are with us. This is our time and our place, and we will remember this day always. We have had some tough times, but we have mastered our work and we are about to make it pay off. You are a hell of a good team, one that I am proud to lead. Whatever happens, I will stand behind every call you will make. We came into this room as a team, and we will leave as a team.

After imparting my final words of encouragement, I get down to business. "All controllers go back to the Flight Director Loop and give me a green." In a matter of seconds, the lights above my television monitors quickly change from amber. I glance briefly at the wall clocks, writing "102+ 12, Battle Short" in my console log. My next words were heard often in training but today took on an altogether new meaning, "Network take Mission Control to *BATTLE SHORT*."

"Roger Flight . . . going to Battle Short." Without a backward glance, George Ojalehto steps from his console to the left door of the control room. Turning the deadbolt on the heavy grey, metal door, he turns, retraces his steps, and walks down a short hall to lock the doors on the right rear of the control room. Now no one can enter or leave the Mission Operations Control Room (MOCR).

It is 2:43 p.m., July 20, 1969, in Houston, Texas. We are going to land on the moon.

The descent and landing phase of Apollo 11 brings me and my team close to the mission abort boundaries many times. My team and I are unaware of three critical challenges we would face. The first, a hardware modification at the Cape involved the LM down-firing thrusters. Their proximity to the LM skin, which is thinner than a sheet of paper, needed to be protected. A scupper was installed—essentially, a curved metal sheet to direct the gas away from the LM. In certain attitude orientations, the metal interfered with the antenna signal sending voice, telemetry, and tracking data. My primary ground rule responsibility through the descent and landing was to "assure that in case of an accident we had adequate telemetry and radar data recorded in Mission Control to determine the cause."

The second challenge was a known but not fixed software interface problem between the LM computer and rendezvous radar. This caused five computer restarts during the descent, distracting Guidance Officer (GUIDO) Bales from a Go/No Go evaluation of landing radar data and Armstrong from an early assessment of the landing site.

The third challenge occurs because the crew had not fully depressed the docking tunnel and the residual pressure accompanied by the effect of LEM reaction jet firings acted as a maneuver and resulted in a three-mile landing point change.

The powered descent consists of three computer-controlled phases. The "braking phase" decreases orbital velocity during descent. The "approach phase" begins at 7,000 feet and continues to about 150 feet altitude followed by the crew takeover for the "landing phase" to touchdown. As the minutes count down to the powered descent Go/No Go decision time, the telemetry, voice, and radar data is intermittent. If I wave off, I will have one more chance to attempt descent. After that, any landing option is off. Even with intermittent data, I determine the crew is in good shape, the trajectory is within limits, and the controllers indicate the LM systems are "Go." The data drops out at my decision time, then we get a few seconds of data. Controllers all rapidly give me a "Go" then data and radar again drops out again. I decide to press on. If data is lost, I have ten minutes to get it back before aborting. At first, the Eagle does not hear our "Go" but we relay the "Go" through Mike Collins in the CSM, and hear the crew confirm the "Go."

The challenges continue. At times, the communications noise is so loud I cannot hear my team. Four minutes into descent at 41,000 feet and 75 miles to touchdown, we give the "Go" to continue descent. The crew rolls to face-up orientation. Now we need landing radar to update the LM computer to the actual altitude above the surface. Armstrong is descending at 90 feet/second, about 60 mph, killing the forward velocity when a loud popping noise again obliterates voice and data.

We give the "Go" to continue powered descent at 35,000 feet, descending 132 feet/second, 50 miles to the landing site. The data seems better, with less noise the crew reports are clearer. Six minutes into descent, the crew are now face-up. They report they see the Earth out the window. Then, both the crew and guidance report "landing radar." Seconds later Aldrin calls "Program Alarm . . . 1202." Then, silence. CapCom asks if we can give them a reading. Bales says, "We are 'Go' on the alarm." We would have five more alarms in the descent, each one causing a computer restart and each one distracting Armstrong who should be looking out the LM window assessing the landing site. Still, like all good pilots, he is ready to take over the LM manual control if needed.

Bales gives me a "Go" on radar-altitude convergence. We now have the actual altitude in the computer. The mission rule is satisfied. My landing team has "IT!" Every controller is making Go/No Go decisions on limited and often stale telemetry and trajectory data, and split-second assessments of LM guidance and control after the computer restarts. My team gives me their "Go's" without me even asking.

At nine minutes in the descent, we give the crew the "Go" for landing. At 2,000 feet, Armstrong views large craters and a boulder field. He decides to overfly the crater. We expect to land with about 120 seconds of fuel remaining but now Carlton, with stopwatch in hand, is calling off the seconds of fuel remaining. 90 seconds . . . 60 seconds . . . and near the 30 second call we hear the crew go through engine shutdown. We are unable to savor the moment. We quickly assess the LM status and make a "Stay" decision. After landing, we continue to assess the LM status for up to two hours before handing over to Charlesworth to prepare for the EVA.

Then we celebrate!

***

Six days after the mission, I received the following memo from Kraft:

> Your performance and that of the flight control team during the landing phase of the mission was one of the most outstanding I have ever seen. I sincerely believe that had it not been for your actions during this period, the landing would have been aborted.

After reading Kraft's letter, I thought of the words of General Douglas McArthur, "You are remembered for the rules you break!"

The succeeding Apollo missions would prove equally challenging. On November 14, 1969, Flight Director Gerry Griffin was launching Apollo 12 with astronauts Conrad, Bean, and Gordon. Griffin, after graduation from Texas A&M in 1956, entered the Air Force as a radar intercept officer on the F-89 Scorpion and early supersonic F-101B Voodoo.

Griffin likes to tell the story how after an interview in 1962, I did not hire him because he wanted too much money. Well, in 1962, I was new to the business and learning the job myself, still trying to figure out how the NASA pay system worked. After a two-year stint at General Dynamics he, "swallowed his pride and his wallet," took the pay cut, and reapplied to NASA in 1964. He considered it the best step he ever made in his life. "There were so few of us, we were all kids making life and death decisions. It was a unique time in a unique setting."[11]

Apollo 12 launch Flight Director Gerry Griffin and Chris Kraft discussing lightning strike

From the beginning, I knew Griffin was a winner and it was moving him alongside Aldrich in Gemini systems that really provided the payoff. Griffin was an Aggie, a weapons systems officer in a supersonic fighter and now a flight controller. In all aspects of his life, he was an "over-achiever" and born team player. In his oral history, he described the young flight control organization this way: "It was as close to being like a fighter pilot organization as I had ever seen. It took a bit of cockiness (measured) and confidence (measured). In both cases, if you did not have the confidence to speak up and get the job done, you would not last long. It just didn't work."

Apollo 12 was Griffin's first mission as a launch Flight Director. A weather front had stalled North of the Cape and was not forecast to move through before launch. Despite the wet weather, the Saturn with Astronauts Pete Conrad, Al Bean, and Dick Gordon lifted off at 11:22 a.m. EST. Thirty-six seconds into flight, all hell broke loose, and Conrad started reading off the caution and warning lights being displayed. Unknown to the crew and Mission Control, the spacecraft had been hit by lightning. Twelve seconds later, it was hit again. Conrad reported, "We just lost the platform. I don't know what happened, everything has dropped out."

The branch chiefs make the controllers' assignments to the missions and when a Flight Director is assigned to a critical phase for the first time, they normally place the most highly-qualified controllers in positions to help the "new flight" out. Fortunately for Griffin, on that day in Mission Control, John Aaron was assigned as his EECOM. Aaron disregarded the many event lights illuminated on his console and intently scanned the TV of the telemetry power displays. With spacecraft power off, John expected to see zeroes on his display, but he saw numbers that did not make sense. *Why?* In a test a year earlier, John had seen the same pattern after a procedural error and traced it to a signal conditioning problem. He immediately advised the Flight Director to have the crew place the Signal Conditioning Equipment (SCE) switch to auxiliary (AUX). No one including the crew knew what SCE to AUX meant. Aaron again repeated, "SCE to AUX." The switch was in front of Al Bean, on the bottom right of the panel adjacent to the tape recorder switches.

When Bean selected "AUX," telemetry was restored. Now with data, Aaron voiced instructions to CapCom Jerry Carr with the crew procedures for restoring power. The incident report indicated that the exhaust plume provided a conduction path to the ground when it entered the field of clouds. Gerry Griffin's first launch was one to remember. Astronauts Conrad and Bean landed on the Ocean of Storms adjacent to the Surveyor 3 spacecraft that had landed on the moon in 1967.

***

The Apollo 13 LM carried the name "Aquarius." On my way to work, I often turned up the volume of the 8-track recorder playing the soundtrack to the musical *Hair* with its song about the ". . . dawning of the age of Aquarius." As I passed security entering NASA, the guards recognized me and if "Moody" was working, he offered a salute and a broad smile. The Apollo 13 story has been told many times and the movie brought Mission Control and the controllers into focus. The movie and most of the books on the mission focused on the event and highlight a few of the key controllers. Throughout those harrowing four days, I never doubted that we would get the crew home because we had "IT!"—the dedicated and superb controllers in Mission Control doing what they knew best and doing it well.

The movie and my book, *Failure Is Not an Option*, did not discuss the operating team structure, the Program Office, Engineering relationships, nor the crisis-management process. Because these were critical to our success, critical to the function of "IT!" I will go into them here.

The Apollo 13 oxygen tank #2 exploded at 55:55 hours into the Apollo 13 mission. Six seconds later, the pressure increased in the service module bay, resulting in the separation of the bay structural cover, and damage to the cryogenic tank valves and plumbing that resulted in the loss of oxygen tank #1. The loss of the cryo tankage resulted in the loss of power output from the three fuel cells. The explosion occurred about 200,000 miles from earth and about 50,000 miles from the moon. Survival of the crew depended on essentially converting the LM into a lifeboat. The basic LM design provided electrical power for two crewmen for two days and the return journey was at least four days with three crewmen. Fortunately, we had the LM power down and docked propulsion maneuver procedures developed during Apollo 9 to support our early actions.

A lead Flight Director is assigned for each mission about eighteen months prior to planned launch to provide mission development oversight, briefings for the media and NASA Headquarters, the focus point for mission rule development, and crisis leadership if needed during a mission. The Flight

Director's assigned team, augmented if necessary, will provide the crisis planning and execution support. The latter is important as the team integrating the plans will support the Flight Director when executing the plan. The Mission Control second floor data room is the predesignated action center for the crisis team.

The Spacecraft Analysis Team (SPAN) is the central point for technical engineering and analysis support for the crisis team. The flight-control division systems branch chiefs, Program Office and NASA and contractor leadership provides the staffing. This room is in operation throughout the mission, tracking mission events and any space system anomalies and has "hot line" communications to the manufacturer and all subcontractors.

SimSup and members of the simulation team are on standby during all mission periods. As problems are recognized, they configure the simulation system to track the "current" configurations of the spacecraft to support procedure development and troubleshooting. The Apollo 13 explosion occurred shortly after a television survey of the LM. After completing the show, the crew returned to the command module and the White Team was preparing to initiate the pre-sleep checklists. Lunney's Black Team was standing by for handover and as Flight Director, he was sitting by me at the console when Will Fenner, the GUIDO, announced, "We've had a Hardware Restart." What happened next was so dramatic, it became one of the more famous lines of American space history: "Houston, we have a problem," is how the country remembers it. In fact, this was the exact exchange:

> 055:55:19 Swigert: Okay, Houston . . .
> 055:55:19 Lovell: . . . Houston . . .
> 055:55:20 Swigert: . . . we've had a problem here. [Pause.]
> 055:55:26 Fenner (GUIDO): FLIGHT, GUIDANCE.
> 055:55:27 Kranz (FLIGHT): Go GUIDANCE.
> 055:55:28 Lousma: This is Houston. Say again, please.
> 055:55:35 Lovell: Houston, we've had a problem. We've had a main B bus undervolt.
> 055:55:42 Lousma: Roger. Main B undervolt.

We had seen Caution and Warning Problems before and my initial thought was that it was an electrical glitch. We will get the crew to sleep and then work the problem. Moments later, Haise reported, "We had a big bang associated with the Caution and Warning." As more reports came, nothing made sense. A crisis had begun, and further reports came in, escalating and complicating the problem. It was taking extra seconds to sort out what was real and credible and tie the crew and controller's reports together. All three crewmen were reporting the meter readings and warning light indications when GNC, Buck Willoughby, calmly said, "Flight have the crew check the Quad D Helium to see if valves are open." Then he added, "I think the big bang shocked them closed." The crew reset the valves.

We were now five minutes into the crisis and Gary Scott, the Instrument and Communications Officer (INCO), became the hero, making callouts for communications antenna switching to protect the vital communications link. Both the crew and the controllers were getting frustrated, and it was noticeably creeping into their voices. Everything we knew about the design of the spacecraft precluded the massive failures we were seeing in the spacecraft systems.

Kraft was at home when I called and reported that we had lost the fuel cells and he should get in quickly. GNC and Will Fenner, the GUIDO, had been watching the spacecraft motions and, along with the CapCom, provided directions to avoid Gimbal Lock. (This is the loss of one degree of freedom in a three-dimensional mechanism that works essentially like the spacecraft's compass, resulting in the inability to properly orient and point.) The team was now functioning well. EECOM Liebergot had tried everything to staunch the hemorrhage of the fuel cell oxygen to no avail, when Lovell called and said, "It looks to me like we are venting something, a gas!"

I should have seen it and caught on earlier. Mission Control was the only hope for crew survival, so I stood up and shouted, "White Team get back on your consoles and on the loops. The rest of you shut up. The LM is attached, we have it to get us home. Let's solve the problem, team," I said. "Let's not make it any worse by guessing." Twenty minutes into the crisis, we started an emergency power down. Lunney and I had a brief discussion, and he went down to work with the trench to develop return-to-Earth options. Within thirty minutes, network communications and data sites were brought up to provide any data they had received from the LM. The crew was progressively isolating and saving all consumable systems required for returning to earth.

Backup computers in the MCC were brought online to support the massive processing of data from the spacecraft recorders. Ed Fendell had arrived from home and joined with Scott to lead the orderly communications transfer to the LM communications.

With Kraft's arrival, Lunney crisply summarized the return options. "We can execute a direct abort coming around the front side of the moon and get home in thirty-four hours. To do this, we will have to jettison the LM and use all the CSM Propulsion system fuel. There are also options to go around the moon that are two days longer, but we can keep the LM, and we don't need to use the CSM propulsion."

I briefly went through the LM status. All consumables were tight, but oxygen looked OK and so was water. Power was the problem as we were about thirty-six hours short.

Fifty-three minutes after the explosion, controllers surrounded my console waiting for a plan to put their teams into motion. Lunney had clearly laid out the options. I remembered the issue I had with Mathews on the Gemini 9 EVA, so I spoke directly to Kraft. "The options of going around the moon buy us time to think. We retain the LM as a lifeboat and to execute maneuvers. We do not need to use the CSM propulsion system. I intend to go around the Moon." Kraft and Lunney nodded in agreement, and upon hearing my words, the guys in the Trench smiled and returned to their consoles issuing the commands to come up with the specific trajectory plan and sending the Navy recovery forces into action. I locked in the plan announcing over the Flight Directors' loop, "All mission planning will be around the Moon!" There would be no more discussion, no equivocation: all program elements, NASA Centers, military facilities, contractors, and the media knew our direction.

Once again, we called upon "IT!" The team chemistry elevated performance at the task and social levels and the team performance increased over the next few hours. The White Team handed over to Lunney's team about one hour after the explosion and assembled with me in the Data Room #210. Lunney's team immediately began the crew transfer process configuring and then evacuating the CSM, ingress and power up of the LM, alignment of LM navigation system, and finally, turning off the CSM computer followed by a total CSM power down.

Four and a half hours after handover, Lunney's team planned and executed a LM propulsion maneuver to place the docked spacecraft on a free return trajectory that, we hoped, would return the crew to Earth with no additional maneuvers.

Meanwhile, with my White Team assembled, my first action was to assign key leadership roles. I told Aldrich to assume the development of all checklists and I gave Aaron the responsibility for all consumable resources on both spacecraft. Bill Peters was assigned to roam between the crisis team elements and the on-console teams to provide oversight—in other words, anything we were missing.

For about three hours, the return maneuver, landing times, and even the ocean landing location was constantly changing. The trajectory team established the return maneuver ignition for pericynthion (closest approach to the Moon) plus two hours (Pc+2) or 79:30 on the clock and provided the Navy recovery forces the target ocean landing point. This decision allowed me to work backward to establish the start of shift time for my team. At that time, we had worked under pressure continuously for over thirty hours, so I asked my flight surgeon to dispense some stimulant pills to keep them focused. I then joined with Arnie and the LM team on the maneuver checklist. The Pc+2 engine procedures were like those we performed during the docked engine testing on Apollo 9, so I was confident in the accuracy of the critical propulsion procedures.

The trajectory carried the spacecraft to an apogee (point in orbit furthest from the Earth) 9,755 miles above the Moon, setting the world's manned absolute altitude record of 248,655 above the Earth. Team shifting worked well and my White Team was back to execute the Pc+2 maneuver that would return the crew back to the mid-Pacific at 142 hours mission elapsed time (MET) hours. Griffin's Gold Team had set us up well. We briefed the crew on the plan and, aware that the crew was tired, repeatedly advised them the maneuver start time for the Pc+2 was not critical.

The LM power down would be to the configuration baseline specified by Aaron. During this sequence, Aaron would be sitting with the LM controllers and, when necessary, advising the CapCom. The best engineering estimate was that the LM cockpit would be near freezing during the return.

Prior to the Pc+2 maneuver, I was briefed by my systems people that the SPAN team was concerned about external temperatures and wanted to do a rotisserie maneuver to warm all sides of the spacecraft in the sun. This action triggered a vigorous discussion at my console with Slayton, the astronaut boss, on crew sleep versus temperature control. My decision was final, the crew would execute a thermal control maneuver perpendicular to the sun line. To solve another problem Ed Smylie's crew systems division developed a procedure to modify the CSM CO_2 cannisters to fit in the LM environmental system to prevent an increase in the CO_2 level. Then we powered down to a twelve-amp load. It was going to get cold.

The second, much more complex task, would involve developing the CSM reentry checklist: the crew transfer from the LM to the CSM, power up of the CSM, an attitude maneuver for for the service module jettison, and LM separation and reentry. At this time, we did not know that a shallowing trajectory would require a LM maneuver to correct the trajectory, aggravating the power profile.

We had several more challenges, including a manual trajectory correction maneuver during the two-and-a-half-day return. The flight operations crisis plan provided the leadership, the team, and the support to save the crew of Apollo 13. Responsibilities were assigned and executed. Priorities were clear. The critical path was established early, with effective resource utilization. Finally, all participants were willing to experiment, innovate, and find solutions. The trust within the teams provided the ability to move rapidly, take advantage of opportunities, and quickly access needed resources with a minimum of discussion. No matter how dire the circumstances, the common belief was that we will get the crew home. There was never any doubt. In the words of Arnie Aldrich, "'IT!' was all built in, we had been working on 'IT!' for years."

***

I had both the privilege and the sad task to write the epitaph of the Apollo program flight operations and close out my career as a Flight Director on Apollo 17. I launched the crew of Gene Cernan, Ron Evans, and Jack Schmitt from Earth and after three days on the surface at Taurus-Littrow launched Cernan and Schmitt from the surface of the Moon. The United States had won the space race and that victory became a turning point for NASA.

13

CHANGE

Our liberty . . . is endangered if we pause for the passing moment, if we rest on our achievements, if we resist the pace of progress. For time and the world do not stand still. Change is the law of life. And those who look only to the past or present are certain to miss the future.

—John Kennedy

What happens to an organization once its primary goal has been achieved? What happens to "IT!" when the mission that fixed our eyes forward is now in the rearview mirror? I was about to find out, and in doing so, face the kind of challenge that all organizations face when they transition from one strategic plan to another.

The Moon's emotional and visceral appeal was absent in the next phase, and I was faced with steering the teams toward low-Earth orbit and the Skylab Program. Skylab was America' first space station. It was derived from unused space systems from the Apollo Program. The "Workshop" utilized the Saturn V third stage to provide habituality capabilities and as the location for biomedical testing. Science provisions included a solar observatory; earth-looking instruments to address land and vegetation patterns, hurricanes, and mineral resources; and instruments to study human physiology and biomedical research.

Skylab was not a JSC priority and with a gap of less than six months, I needed to quickly convert my Apollo teams to support at least a year of continuous Skylab Operations.

JSC Center Director Dr. Gilruth retired, and Kraft became the JSC Center Director. He selected Sjoberg as his deputy and Bill Tindall as the Director of Flight Operations. Kraft and Sjoberg had watched my teams perform throughout Apollo and were fully knowledgeable of the talent pool in my division. For more than a decade, Mission Control served as the training ground for developing young, capable leaders with the ability to take on NASA's challenging missions. Kraft and Sjoberg knew I had leadership depth and that I could develop the needed operations leadership for the Skylab Program. Kraft made several moves: Arnie Aldrich was assigned as Deputy Project Manager to Kleinknecht for Skylab, Gerry Griffin was assigned to the NASA Headquarters Office of Legislative Affairs, Cliff Charlesworth was assigned as Chief of the Earth Resources Project Office, and Glynn Lunney assumed the responsibility as Technical Director of the Apollo Soyuz Test Program (ASTP). While it placed a load on my division to fill the vacancies, it had the beneficial effect of providing the incentive to develop the next generation of NASA operational leadership.

Kraft, in a further reorganization of JSC, assigned the Landing and Recovery Division function and personnel to my organization. With only the Skylab and the U.S.- Russian (ASTP) programs requiring crew and spacecraft recovery in the ocean, I phased out the Landing and Recovery Division and established Mission Recovery Support as a section of the Flight Control Operations Branch. Other recovery personnel were assigned to form an Aircraft Applications Branch, supporting Earth Resources flight operations in various roles as project managers, mission managers, and data systems engineers.

The first step in addressing the post-lunar challenge was to recognize that a letdown is inevitable after a major change in mission content. The key is to anticipate the letdown and identify those elements most affected. Therefore, my primary approach as Apollo ended was to research other areas of the Center where I could relocate my personnel for one to three years to satisfy temporary critical needs and then recover them in time for shuttle mission preparation. The transition from the Moon to low-Earth orbit had a significant impact on the trajectory team, the "Trench." During Gemini and Apollo, the Trench wrote the textbooks for exploring space through the applications of the laws of physics and orbital mechanics. Their trajectory work was at the core of every mission—exciting, challenging, and demanding. Now, the Trench's work in Skylab centered on a few launches, a spacecraft orbiting the earth endlessly, and finally reentry.

Retrofire Officer John Llewellyn accepted the challenge of change and rocketed into his new challenge in support of the Large Area Crop Inventory Experiment (LACIE), a Skylab venture that used satellite technology to analyze crop yields, helping to forecast global production. This surprised me because Llewellyn was the hard-core Retrofire Officer for the final Mercury missions, and most of Gemini and Apollo. First and foremost, John was a Marine artilleryman who served with the 1st Division in Korea in the drive North to the area of the Chosin Reservoir. When the Chinese crossed the Yalu River, all units in the Chosin area were engaged in a brutal, close battle for survival, fighting seven Chinese divisions. The withdrawal from the Chosin Reservoir is revered as a high point in Marine Corps history. After the war, John entered William and Mary College, received a degree in physics, and subsequently joined the NASA Langley Research Center.

My first contact with John came when he was assigned to the Mercury remote sites. After observing his mission performance through early Mercury, Kraft reassigned him as a Retrofire Officer in Mercury Control. John knew I had served in Korea, so when drinking beer after Judo or on other social occasions, I became the go-to guy when John needed to "power down."

John bought property in Alvin, Texas, where I met his wife Olga, who was John's perfect match. One of the favorite Llewellyn stories is when he overslept and arrived late during Gemini 7/6, he was unable to find a parking place, and he just drove up the steps and parked his car at the entrance to Mission Control. Security pulled his permit and the ever-resourceful Marine brought his horse trailer, parked it on the highway, and rode into work, tying his horse to the bicycle rack at Mission Control. John was the carbon copy for the "Captain Refsmat" poster described earlier.

John never did anything halfway. His orbital trajectory skills were well utilized mapping worldwide ground targets for crop studies and setting up the Skylab instruments to collect data. His zest for the work established strong relationships with the Earth Resources Principal Investigators in Mission Control and at the test sites on the ground that had been extensively documented for crop and soil comparison purposes ("truth sites"). The survey data obtained was so good that during many Skylab data passes, the data received in Mission Control was embargoed to prevent financial crop speculation.

The Skylab Earth Resources Program utilized the NASA JSC sensor-equipped aircraft to obtain qualitative measurements to support the testing and calibration of the spaceborne Earth resources sensors. Three aircraft—a B-57, C-130, and an old Navy P2V-7 Neptune—were outfitted with sensors requiring airborne operators. Dutch von Enrenfried and several others volunteered for aircraft operations and were welcomed by Joe Algranti, the Chief of NASA Aircraft Operations to support the sensor engineering, installation, checkout, and airborne operations. Following the Earth resources program, I planned to recover the engineers in time to support the Shuttle approach and landing flight test (ALT). My final move was assigning Hannigan's lunar module avionics, power, and propulsion section personnel to establish a foothold in the Shuttle Program.

At NASA Headquarters, Bill Schneider held the overall program management and integration responsibility for Skylab. Schneider was a Navy aviation machinists' mate and petty officer 1st class. He was the Headquarters Mission Director during Gemini and early Apollo and we had worked well together. I believe his Navy work provided a solid feel and a "get the job done" approach to work in Mission Control and as a leader, he was well equipped to control the politics for Skylab with the work distributed across many NASA locations. The major Skylab module systems were developed by the George C. Marshall Space Flight Center (MSFC) in Huntsville, Alabama. My Apollo systems engineers transitioned easily in preparation to fly Skylab and, on an engineer-to-engineer basis, developed solid interfaces with the engineering and program personnel at MSFC.

My division's Skylab work schedule, to a great extent, was directed by the MSFC Program Office. The Space Shuttle was the next upcoming major program for JSC and, as a result, Skylab did not rank high for personnel, budget, travel, overtime pay, or logistic support. As I fought the early battles on colocation of NASA and contractor engineers, the battle now swung to issues over center policies and support for a program that was not managed by JSC. I was a "wild card" in JSC management. I visualized Skylab as a laboratory prototype of future space systems. During reviews at MSFC, I recognized that Skylab was a vastly different space system that would challenge my controllers and provide the "lessons learned" for future orbital space programs.

My experience in the Air Force and on the B-52 was, "If you are not growing, you are marching in place."

Change taught me to learn new systems, adjust operations concepts, and meet and work with different engineering and management teams.

Skylab was in effect the first space station. It consisted of four functional modules controlled by the Instrument Unit computer used on the Apollo Saturn rocket and its mission management would provide a significant challenge to my guidance system controllers. The new systems were key to long-duration earth orbital spaceflight: solar power, control moment gyros for attitude and pointing, massive communications downlink processing, teletype printer uplink, leak-detection techniques, and in-flight maintenance. Once these challenges were overcome, we would possess operational experience with a large space system to serve as an engineering and operational testing ground for future long-duration space systems.

While we were preparing for Skylab, Hollywood revitalized the science-fiction genre for the movie screen. With the pending launch of an American Space Station, science fiction preoccupied the movies. In 1972, a decommissioned aircraft carrier, the USS Valley Forge from Korea, served as the interior of a space freighter for the movie *Silent Running*. Starring Bruce Dern, the film imagines a time when all plant life has become extinct. To save various endangered species, environmentalists created gigantic dome-like greenhouses which they attached to spaceships in anticipation of a day when they can return to Earth and begin the process of re-forestation. Dern's character is a conservationist who rebels when, in order to save costs, he is ordered to destroy the plant life on the station. The movie was my favorite for many reasons. It was a single man challenging the system; a needy, lonesome man, forming a relationship with a drone. It made a journey beyond the realm of Earth come alive. I occasionally had the same feeling, night flying over the Pacific amid the stars, alone on a journey.

While I looked forward to the Skylab challenge, I knew that Skylab, like *Silent Running*, would not provide the emotional content of Apollo and I would be hard pressed to motivate my organization to prepare for a program of continuous operations lasting over a year. The challenge was mine and the branch chiefs, but I had to set the example for the division.

To save "IT!" in Flight Control, I must demonstrate the same passion and enthusiasm that I carried during Apollo. I decided to literally "dive" into the Skylab systems and the mission science. I learned the interface with the scientists, conducted mission rules reviews, and provided the JSC operations leadership face to Headquarters, Kleinknecht, Aldrich, and the JSC Skylab managers. Other than the involved paperwork, organizational charts, contract statements, and budgets, the transition of flight operations was easy. The branch chiefs had sensed the opportunity, joined in, and their passion soon carried throughout their organizations.

Skylab was a large space system that utilized new, long-duration technologies. The science objectives were broad based and provided opportunities to learn alongside the scientists. I sent two controllers to one year of pre-med to establish the background for the space medical experiments. (Unfortunately, only one returned; the other gave up on space, and reimbursed NASA for his medical training and completed medical school.) The long duration offered challenges for planning, training, systems management, and daily operations. Five full mission teams were required: three teams for daily operations, one team for planning the subsequent mission, and the equivalent of one team for office work. All teams rotated through the Mission Control shift schedule, and virtually every member of operations was a mission controller providing "bragging rights" so they could say "I work in Mission Control."

Secretaries were trained as teleprinter controllers (TPCOs) to prepare and send flight plans, procedures, and status messages directly to a machine onboard the workshop. A full-time printing press was installed in the MCC to copy, publish, and distribute daily flight plans and update mission rules and manuals to accommodate the constantly changing nature of the program. The Mission Control teams adjusted readily and rapidly to the different nature of the work. We had successfully transitioned to the post-Apollo programs, conveying "IT!" to our new teammates from other NASA centers.

14

SAVING SKYLAB

Innovation is taking two things that already exist and putting them together in a new way.

—*Tom Freston*

Skylab was the program that established America's first space station. The program had four classes of objectives: earth resource observation studies, advanced solar and nearby star studies, studies of the Zero-G effects on man and other living organisms, and a broad range of corollary experiments. The Skylab spacecraft consisted of four major elements: a Workshop Module derived from a Saturn IVB upper stage that provided living quarters and a space laboratory with two large deployable solar arrays attached to the external structure; an Airlock Module supported EVA egress recovery of film; a multiple docking adapter accommodated CSM docking and finally, an Apollo Telescope Mount supported five major solar telescope instruments

The flight program consisted of four missions—an unmanned mission to launch the Skylab station followed by three manned missions. The subsequent missions of three crewmen each provided increasing durations of one, two, and three months with one-month unmanned intervals between missions. Extensive ground command controls provided the ability to control the Skylab and operate experiments between the manned periods.

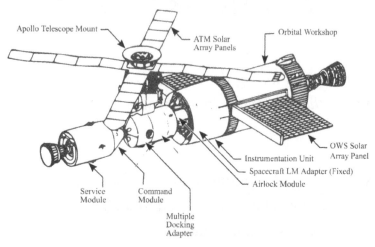

SKYLAB schematic

1973

Because of the Apollo schedule, my mission operations support was a late arrival to the Skylab program. After Apollo 15, Milt Windler was assigned as Lead Flight Director for Skylab, Don Puddy was assigned to the Skylab on-orbit activation, and Philip C. Shaffer, with a trajectory background, would launch all manned missions.

The unmanned Skylab 1 mission was launched atop a Saturn V rocket on May 14, 1973. As the Saturn V rose majestically, launch responsibility was handed to Flight

Director Don Puddy at "Tower Clear." Sixty-three seconds into flight, at the point of maximum dynamic pressure, a structural failure occurred and telemetry indicated micrometeoroid shield (MS) deployment and solar-array beam fairing deployment. The launch, however, continued apparently normally and after attaining the planned orbit, the orbital activation sequence was initiated. The radiator shield and payload shroud were jettisoned, and the Apollo Telescope Mount (ATM) rotated to the flight position above the docking module. Skylab was now configured to allow CSM docking. The ATM solar arrays were extended followed by a maneuver to place the Skylab axis in the orbit plane with the ATM pointing to the sun. The fight to save Skylab began now and never again was there a dull day in Skylab Mission Control.

At the Canary Island tracking station, there was no indication of the Micrometeoroid Shield (MS) and solar-array deployment. Subsequent stations indicated that temperatures on the exterior of the workshop were rising rapidly. The MS provided a secondary function by providing "shading" of the workshop main structure. The immediate concern was high temperature outgassing of toxic materials into the station's atmosphere and damage to the onboard film and food. Experienced in crisis operations, the mission teams assumed manual attitude control of Skylab and for the next six days, via ground command, maintained an orientation utilizing measurements from external workshop skin temperatures that provided adequate solar-array power and minimized

OWS interior temperatures. This approach used the control moment gyros augmented by orbital nitrogen fuel.

The controller team structure had an immediate payoff. The systems controllers were trained to address problems in an integrated fashion. The two immediate problems, power and thermal, were the responsibility of a single controller. The gyros and attitude control were the responsibility of another controller position. Mission Control's experience in crisis management assumed the leading role to stabilize the situation and, like Apollo 13, quickly executed needed attitude control actions balancing power needs and internal workshop temperature control. Pete Frank and I were serving as Flight Operations Directors on a twelve-hour shift schedule.

Saving Skylab depended on the Mission Control team keeping Skylab alive while the MSFC and JSC engineering teams developed schemes to shield the exposed workshop skin to control the internal temperatures and deploy the remaining OWS solar-array, which our data indicated was jammed closed. Outside the Mission Control room, there was a feeling the program was lost as the impact of the launch structural failure was evaluated by the Program team during meetings. The mood became visible to the media on the first day when the normal change of shift press briefing was moved earlier, and the general tone was pessimistic.

The controllers, well experienced in crisis management, were working with the MSFC engineering to optimize the management of the Control Moment Gyros (CMG's) to maintain attitude orientation with minimum use of the attitude control nitrogen fuel. During the pre-mission Systems/ Operations Compatibility Assessment Review (SOCAR) stage, MSFC and JSC slugged out design and operational issues. In the process, both groups developed a deep, mutual respect. This was the moment when I felt the "IT!" team chemistry emerge, giving Skylab something of the "Tough and Competent" ethic of Apollo. The Skylab Mission Control operations team structure provided integrated systems monitoring and control across the four modular elements: Orbital Workshop (OWS), Airlock Module (ALM), Multiple Docking Adapter (MDA), and Apollo Telescope Mount (ATM).

My deputy, Mel Brooks, had established a close relationship with MSFC/George Hardy and JSC/ Reg Machell during the SOCAR. This marriage of convenience now became one to save Skylab. It provided lines of communications that bypassed the NASA center prerogatives and allowed total focus on the actions and responsibilities needed to salvage Skylab. At the top-level, George Hardy and I literally assumed the leadership responsibility to address and resolve all issues for our respective centers. Temperature control was the immediate challenge, but it had to be balanced with establishing an orientation to illuminate the solar array and produce power to charge batteries for use during the orbit night cycle. Pete Frank and I were tasked to plot trends for the team, as well as keeping management appraised of the status. Real-time experimentation determined that the best attitude to provide power and minimize internal OWS temperature was a 45-degree sun angle. To establish the OWS roll, orientation external temperature telemetry measurements on opposite sides of the workshop were balanced, while the pitch attitude was determined by electrical output from the solar arrays. Much of the MSFC-provided engineering data was metric while Mission Control worked in English units. In a day before small, hand-held calculators, I quickly returned to my slide rule for rapid conversion.

The ground team with "IT!" took over control of Skylab and for the next ten days, bought time for the astronauts and engineering teams at JSC, MSFC, and Langley to develop and test fixes to save the Program.

MSFC established the "survival" attitude orientation that established an OWS internal temperature of about 130 degrees with 2800-watt power output that would save critical experiment, film, and food in the workshop. Ingenuity and perseverance saved the $2.5 billion program in a race against time, which was described in several papers as "10 Days in May." Originally, I had planned to utilize four teams for SL-1, providing for three shift coverage with the fourth team planning the subsequent mission. This scheme quickly fell apart, so we moved to two, twelve-hour shifts in Mission Control with the other teams working the many action items.

The controllers who had saved Apollo 13 now launched into the Skylab salvage effort with the same passion and "never-surrender" attitude needed to save Skylab. "IT!"— the team chemistry—was centered on developing concepts, triggering options, and integrating with the crews at various locations and work groups at JSC and MSFC. During Apollo 13, I was able to centralize the effort into a primary location. The Skylab effort initially did not seem to have a single-focused effort. Throughout the entire NASA organization, there were many action groups formed to address the near term, save Skylab actions.

For the first two days, I only witnessed the Mission Control effort to fly the workshop from the ground, keep it alive, and control the temperatures. We were one small effort, but others were assessing temperature solutions ranging from spray painting the OWS to a series of sunshades on poles, tents hung from the ATM, inflatable curtains, or extending "some" device through the workshop experiment airlock. A Stand-Up EVA (SEVA) to install a device from the open cockpit of the CSM, initially seemed to be the most practical solution but was quickly dropped. Astronaut teams were assessing proposals, but with launch windows on a five-day cycle, time was running out as the launch day was closing in, so simplicity was essential. Fabrication, procedures, training, and stowage were now a priority and a decision on the sunshade concept was needed soon. Mission Control was expending the workshop thruster attitude control system (TACS) nitrogen control gas and that would soon become a program lifetime driving function.

In Houston, Jack Kinzler, a journeyman toolmaker and head of the JSC Technical Services Division, was moving down a different track. While everyone was working on rigging the thermal cover from the outside during an EVA, he asked, "Why not do it from the inside?" He walked over and climbed inside the Skylab trainer in building eight, saw the experiment airlock on the OWS side where the thermal shield had been lost, and then returned to his office. There he wrote out a purchase request for four extendible fiberglass fishing poles. He then sketched out a hub to attach springs to the bottom of the fishing poles, got a twenty-four-foot square piece of silk from the parachute shop and attached the silk and a line to the tips of the fishing rods fabricating an upside-down umbrella.

He hoisted the device on a crane above the shop floor, pulled on the rope, and the assembly deployed very much like an umbrella. Over the next six days, the concept took flight form as mylar (.0030 in. thick) replaced the parachute silk. The fishing poles were replaced by aluminum rods, and the coiled springs were replaced by rat-trap springs. A twenty-eight-foot metal center pole was fabricated from thirty-six-inch-long threaded tubular pieces which all had to fit into an eight-inch square, thirty-six-inch-long metal container. When the scheme went from concept to completion the umbrella was designed, built, tested, and selected over all the others for its simplicity. Kinzler was awarded the NASA Distinguished Service Medal for his invention saving Skylab.

An elite organization in motion addressing a true crisis is a marvelous entity to watch. Throughout my professional life, I had been in the center of the action and now as the Flight Operations Director (FOD), I was more an observer of the action, watching the invention, and the team chemistry of the many elements converging toward the goal to save Skylab. As a Flight Director, I was often at the real-time action center, but now I had the opportunity to stand back and watch the marvelous and complex interaction of the trajectory, systems, planning, communications, and leadership at all levels of JSC operations working with George Hardy's MSFC team.

The controllers all possessed superb technical abilities and genius, but their power came from the integration of their individual and collective chemistry that marked them as *elite*. That was the motivation that Kraft, Hodge, Lunney and others like me, created—the urge to belong, to share a purpose, and to be an integral part of something great during our years in mission operations.

With the umbrella concept moving forward, the next challenge was increasing Skylab's power. Telemetry analysis and some vague imagery indicated a partial deploy of the remaining solar array. This was an important discovery, as the long-term Skylab use depended on power and the remaining workshop array was key. Concepts for the panel release converged on a SEVA with a crewman standing in the hatch while soft docked, using a pole pruner with a scissor-like cutter to release the jammed array. At this point, Mission Control now joined in with the astronauts, crew systems, and engineering to fully develop the concept, tools, and procedures. During this period, the SL-2 crew was still in quarantine, but they and the backup crew were testing, evaluating, and training on all the various thermal protection schemes in one moment, then swinging over to solar-array deployment in the next, while we all watched the clock countdown to the next launch day opportunity.

While power and thermal control were top priorities, Skylab attitude control problems continued to vex Mission Control. Three twenty-one-inch gyroscopes, weighing 155 pounds and spinning at 8950 rpm, produced the torque that was used to control orientation. The system used sun and star trackers and other smaller-rate gyros to determine orientation. The rate gyros drifted over a period of time and the ground had to compute drift rate and issue commands to correct orientation. This was our first experience with this system, and it was essential to minimize use of the nitrogen system for attitude control. The MSFC team provided the control-law expertise to guide the controllers to address the ever-changing challenges of controlling Skylab orientation. "IT!" continued to thrive with every day's victory keeping Skylab alive. Some were small victories, often as simple as developing a new workaround, while others came from MSFC establishing In Flight Maintenance (IFM) modifications to work around problems by replacing bad gyros on the subsequent mission. Each day we kept Skylab alive was a victory!

Skylab 2 with Commander Pete Conrad, Science Pilot Joe Kerwin, and Pilot Paul Weitz was launched May 25, 1973.

Skylab 2 Crew. Joe Kerwin, Pete Conrad and Paul Weitz

Stowed under the seats and in various locations were brown rope, various parts of an umbrella sunshade, pruning pole, and other assorted gadgets. After rendezvous on the fifth orbit, the crew performed a fly-around inspection and during station keeping, provided a detailed verbal description augmented by television of the Skylab configuration. Conrad soft docked with the capture latches, the cabin was depressurized, and Weitz was held by Kerwin, by the legs, as he tried to pull the solar array free with the Shepherds hook. It didn't work—the laws of physics applied and each pull brought the two spacecraft together amid attitude control jet firings. Docking was the next step, but the capture latches did not work. The fallback plan required another depressurization with Kerwin crawling into the docking tunnel to remove the hatch and cut wires to bypass the capture latch relay, then re-install the tunnel hatch. Conrad then redocked, thrusting hard against the docking ring to collapse the probe and hopefully, fire the latches and achieve a "hard" dock.

The crew entered the Skylab orbital workshop on the second day, sampled the OWS atmosphere and with a stifling hot environment of about 125 degrees, began working in fifteen-minute increments. With periodic rest and rehydration over a five-hour period, they deployed the parasol that slowly spread out to cover the workshop skin. Over the next few days, the temperature dropped to 86 degrees as the crew continued activation. Power limitations resulted in limited scientific work, so it was essential to deploy the jammed solar array. The backup crew, led by astronauts Schweickart and Musgrave, had been working on an EVA plan in the MSFC water tank. There were no handholds, no footholds, and no visual aids or lights. It was going to be tough and risky.

I listened to Musgrave read the plan to the crew. I remembered the Apollo 9 EVA with astronaut Cernan when the lack of restraints and tools were combined with a complex array of tasks that resulted in aborting the EVA. This EVA would be extremely difficult with a very high risk that was essential to saving Skylab. With an EVA plan in place, the TPCO's now joined the controller ranks to save Skylab, typing and transmitting many pages of instructions and procedures to the crew that would be reviewed with CapCom Rusty Schweickart the next day. After a rehearsal in the OWS, the crew ventured outside the airlock on June 7. The EVA consisted of assembling six rods into a twenty-five-foot pole with a cutter and pull rope on the end, to cut halfway into the strap restraining the solar array. At first this failed, because without any handholds Kerwin could not stabilize the pole. During the night pass they discovered a U-bolt on the workshop surface. Kerwin used a spare tether to pass through the U-bolt and attached both ends to his suit. The tether and his legs provided perfect stability, and the cutter was attached in a few minutes. Then, with Kerwin holding on to the other end of the pole, Conrad was provided an EVA path to attach a rope with hooks into a ventilation opening of the jammed solar array. Kerwin then tied the other end of the rope to a beam on the workshop exterior. This "kluge" was designed to break the strap restraining the solar array deployment.

The EVA crewmen then squatted, placed the tether over their shoulders, and then stood upright, breaking the hinge loose. When the hinge broke loose, the solar array panel swung outward into position. However, without foot restraints, Conrad and Kerwin were catapulted, "ass over teakettle into outer space," restrained only by the length of their tethers. They then went hand-over-hand to find a secure location on the workshop exterior and looked aft. The solar panel cover was fully erect, and the panels were slowly coming out as they warmed in the sun. Touchdown! Solar heating slowly finished deploying the array generating 7 KWh of desperately needed power. Now with power to support the many Skylab systems and with the team experience to initiate full operations, the door opened to support the program's scientific objectives. EECOM John Aaron described the first month of Skylab as a salvage operation. "We literally flew that thing by the seat of our pants for about four or five weeks in a mode that it had never been designed to fly before. It was a multi-day intensive, creative fly-by-the-seat-of-your-pants problem."[12]

Skylab 2 with the famous "umbrella fix"

Skylab 2 set space records: the longest duration crewed spaceflight, the greatest distance traveled, and the greatest mass docked in space. Two more missions were flown of increasing duration with the four controller teams augmented by a fifth team composed of the mission simulation instructors. The controller teams, who thought Skylab would be boring, quickly changed their minds. From my standpoint, Skylab provided the real-time experience to gain the knowledge to design and operate large and complex space systems like those that would be designed and operated in later decades.

Skylab set precedents for the concept of inflight maintenance, ground control of experiments, 24/7 multi-science operations, and new scientific discoveries were made. There was a constant evolution of new procedures, workarounds, and design fixes for the first-of-a-kind, large-scale, Earth-orbital space system. The attitude orientation and pointing capability of Skylab was provided by three, twenty-one-inch control moment gyros, rotating at about 9,000 rpm which eliminated the need to use expendable gas thrusters for maneuvers. When one of the CMG's failed, the program was again at risk, but MSFC anticipating a CMG failure had a software workaround to provide control with two CMG's. The second failure was a major short in one of the forward power distribution assemblies requiring significant troubleshooting to identify areas affected and then modify checklists and develop workarounds. This was the largest manned space system ever deployed on orbit and each of the missions provided new American space records for mission duration and science achievement.

Over the course of the three missions, most of the crew and experiment procedures were rewritten several times based on lessons learned, because of increased crew proficiency, or system and experiment workarounds. Experiment data recovery frequently saturated the Mission

Control data processing system. Lead times to assess the data and provide feedback into the flight plan were established. Each of the mission teams developed their own unique identity, ranging from team colors to wakeup music selection, baking contests, and team events. I found it is wise to leave the "morale building" in the hands of the leaders of the effected organizational elements. Lacking parties at the Singing Wheel or Hofbraugarten, the Flight Directors often celebrated at their homes with their team. Every Skylab team had "IT plus!"

There was a hiccup in the first weeks of the third mission. With the success of the first two flights, and the realization that there would be no more chances for crewed long-duration space research for many years, the Program Office was inundated with new proposals. They accepted many of them but a lesson learned on the first flight, that astronauts need a week or more to achieve full speed operating in weightlessness was overlooked. The timeline was too crowded. Gerry Carr and his crew fell behind, and it took a few weeks for crew and schedulers to optimize the schedules. They finished by overachieving, and returned healthier than the first two crews, due to a modified diet and more exercise.

Skylab showed that humans could live a healthy life for ninety days doing useful work in space. It gave NASA the confidence to proceed, in the 1990s, to build a Space Station.

***

Continuous operations are a great training ground. New controllers were able to gain as much real-time experience during the initial Skylab mission as many veterans had acquired during all of Apollo. To a great extent, the impact of all employees working 24/7 on a single mission created an organizational chemistry within and between the mission teams. The mission workload and 24/7 schedule involved support teams that incorporated the administration and secretarial elements to work side-by-side with the controllers writing Skylab history as they moved through each day's activities, amending procedures and documenting workarounds.

Skylab orbited the earth 2,796 times during the 171 manned mission days. The crews performed ten EVA's, supported 2,000 hours in experiments, and obtained 127,000 frames of film of the Sun and 46,000 frames of the Earth. The experiments recorded solar flares and confirmed the existence of the Sun's coronal holes.

The Skylab operations period was successful because of the relationship between Bill Schneider's Headquarters Program, the MSFC Program Office, Engineering and Operations, and the JSC Mission Operations and Engineering elements. During prior programs, MSFC and JSC had a reputation that they were unable to work together. We proved that when the top brass stepped aside, the operations and engineering elements came together with the chemistry, the "IT!" to save the program. At the completion of every program, all organizations summarized the lessons learned of benefit to future programs. During Skylab operations, the majority of JSC management were totally preoccupied with the advent of the Shuttle Program.

There were many lessons pertinent to the Shuttle and, eventually the Space Station, which should have shaped future program policies and configuration decisions for the Shuttle and the International Space Station. The most important lesson learned during Skylab was the extraordinarily successful multi-center working relationships that developed internally at Mission Control that saved the program. During the mission, we recognized and successfully resolved several "roles and missions" issues at the working level. The incentive was that if we failed during a mission, the Mission Operations elements at JSC and MSFC would carry the burden for failure. This was a critical lesson to carry into the future manned space efforts.

15

PROJECT MANAGEMENT

Every project is an opportunity to learn, to figure out problems and challenges, to invent and reinvent
—David Rockwell

The last Apollo mission was completed in 1972 when we shifted our attention to Skylab and the Space Shuttle, the world's first winged, reusable spacecraft. How does an organization make a transition from one history defining project to the next? How does it make the transition from one set of leaders to a new generation of leaders? For us, this happened all at once. The lesson was, we do not work in a vacuum. Politics played a significant role in space exploration. When the Cold War demanded we show the world our technological superiority, there was enthusiasm for NASA. When the economic and cultural woes of the 1970s demanded that we address problems at home, there was less budgetary flexibility for NASA's expensive endeavors and in years to come, political, security, and financial objectives would marry the Space Station program for better or for worse to our former Russian adversary.

"Behind the power curve" is an aviation expression that refers to the point in flight, usually either coming in for landing or when rapidly slowing down to lose altitude, when the airplane's drag starts to slow it down faster than the engine can recover quickly. In this case, you can increase the power on the engine, but it will take a while, or maybe never, to build up the engine power to again accelerate the airplane. The aviation expression has been generalized to refer to situations where you are cognitively behind external circumstances and it will take a lot of catching up to become fully aware of a developing situation or event. In her book, *Intangibles,* Joan Ryan states, "No team wins without task chemistry. But almost all championship teams have social chemistry---the bonding, trust and caring.[13] Sustaining "IT!" in an organization demands leadership that senses change and assures that their organization is consistently ahead of the game, knowledgeable, and well prepared for the work to be done. At this stage in my career, like the post Apollo transition, I once again had to dive in to acquire Project Management skills and fortunately, I had two very capable mentors.

1974-1982

Change was the name of the game as NASA began adapting to the needs of longer duration, cost-effective programs in near-Earth orbit. With the completion of the Skylab Program, consolidation of the JSC Operations Divisions and support elements was initiated. Kenneth Kleinknecht moved from his Skylab program role to assignment as Director of Flight Operations and I was assigned as his Deputy Director. The Astronaut Office, Aircraft Operations Division, Crew Training and Procedures Division, Flight Control Division, Aircraft Engineering Division, and Shuttle Training Aircraft (STA) flight test were now integrated into a single organization.

The Space Shuttle program was the fourth human spacecraft program carried out by NASA. The purpose was to materially reduce the costs of taking payloads to orbit by a reusable space system, with a high flight rate that provided accommodations for U.S., foreign, and military users. The shuttle was the size of a small eighty-five-ton commercial jet transport capable of a dead stick (unpowered) runway landing.

The landing approach was three times steeper than a conventional aircraft and astronaut training required the development of a shuttle training aircraft (STA), a modification of a Grumman Gulfstream II, to provide a real-world simulation of the shuttle landing profile. The Program Manager of the STA flight test was Charlie Haines, a graduate of Texas A&M, who obtained an ROTC assignment to Edwards AFB upon graduation.

The diversity of the work between shuttle preparation, STA flight test, and aircraft operations demanded every leadership skill developed in my early career. I mentally thanked Harry Carroll and Ralph Saylor for allowing me opportunities to gain "hands-on" startup experience for new test programs. The response of the new directorate led to a very enjoyable period in my career. With the STA and shuttle, we were moving to do once again something that had never been done before.

Director of Mission Operations, Kenneth Kleinknecht

Often at the end of the day, I was back in the realm of learning engineering from Kleinknecht while teaching him flight operations. After graduation from Purdue in 1942, Kenny worked at NACA Lewis Research Center before serving as advanced projects manager at Edwards AFB. He followed the X-15 Program from inception through first flights. He served as the Mercury Program Manager and served in deputy program leadership roles for Gemini and Apollo. Kleinknecht was thin, tall, and lanky, like a scarecrow. He wore old-fashioned, round glasses, was 100 percent engineer, and he excelled in discussions on design, manufacturing, costs, and schedule judgment. He was not an organized notetaker, always searching for writing materials that he shoved into his pockets, and during meetings he spoke with a raspy, baritone voice. Born in 1919, Kenny reminded me of my college teachers who were deep into math and engineering. He spent most of his time working with the manufacture's design and management team and was seldom involved in operations requirements, mission content, planning, and with the astronauts. With his great technical background, he and I were a good fit, but his leadership style was not a good match for the work, personnel diversity, and intensity of flight operations. He normally expressed opinions behind closed doors, except when he was angry and then he would go on a roll.

Many organizations have internal groups or cliques. Within the Astronaut Office, many believed that George Abbey, Kraft's Technical Assistant, had the deciding vote on crew assignments, media roles, and aircraft utilization among other things. The cliques worked outside the established working groups, generally at the factory, generating space system change inputs that I was forced to support or reject in the formal change board meetings. During the early years of space, astronauts served in many positions as controllers in Mission Control and throughout the mission preparation process. After the Mission Operations Directorate (MOD) reorganization, I quickly got to know and enjoy working with the aircraft operations, flight safety, and crew support personnel. There were elements of the Astronaut Office, however, who believed that Kleinknecht was incapable of providing the needed political leadership of operations.

On October 2, 1974, I provided Kraft a briefing on the Shuttle Carrier Aircraft status and the results of an audit he requested on aircraft operations. When I was the Flight Control Division Chief, I published an annual "State of the Union" that provided performance, resource utilization, and plans to meet the coming years' flight program. On October 31, six months after the consolidation, I published the first "State of the Union" for the consolidated Flight Operations organization. The State of the Union memo addressed interfaces between elements, training costs per flight, supervisors not stepping up to new roles, lack of cross training within organizations, travel costs to support questionable outside activities, aircraft utilization, material inventory, and related work. The memo concluded with, "I will initiate actions soon to better support future JSC Programs. This will require difficult decisions, concerted actions, and organizational change that will affect major elements that may be personally distasteful."

Kleinknecht had a vast experience in addressing and leading space systems development programs, but limited knowledge of operations. Within the consolidated organization, he gave me "carte blanche" to assess performance and to plan and execute needed organizational changes. Unfortunately, my direct approach to issues in briefing Kraft and in the State of the Union rubbed the astronaut corps and aircraft operations the wrong way and concerns were carried to Kraft through George Abbey. Abbey was Kleinknecht's polar opposite. He had worked in many areas of the manned space program for a decade, mostly as a staffer in the upper levels of management and the Program Office where he developed a good reputation. His principal specialty was direct oversight of the astronauts, where he developed acolytes to support his decisions, whatever they might be. George had done good work after the Apollo 1 fire as a staff man for George Low, serving as the early technical secretary for the Apollo change control process. Kraft considered Abbey a top-notch leader. Abbey personally reminded me of staffers I had met in the military, knowing the right people, and punching the right buttons to move ahead.

I had worked for Kleinknecht for two years as MOD deputy director when, in January 1976, I was advised by Kraft that Abbey was moving in as director. Most of my former division supervisors believed this would result in immediate and sustained conflict. They were wrong, regardless of our different personalities. George and I worked rather well together. He addressed the needs of the astronauts and the Aircraft Operations Division and as the deputy, I worked as the Deputy Director with the Mission Control and training elements. The primary area of conflict occurred during the preparation for the shuttle program involved the Mission Operations Director (MOD) function in Mission Control. Abbey believed the way to make decisions was to build consensus by touching base with those he knew and seeking their opinions to form a consensus on direction. I considered the Flight Director as the mission decision authority and the MOD was the Flight Director's "blocking back." I was using Kraft's Apollo role as the model:

> During Mission Operations periods, the Mission Operations Director has the responsibility for overall assessment of mission support associated with the adequacy of training, planning, and pre-mission operations readiness. During Mission Operations, the FOD provides the interface between NASA management and the Flight Director on the topic of overall mission strategy.

As Deputy for Mission Operations, my favorite work was representing the Astronaut Office and the Mission Control team at the Orbiter Change Control Board (CCB). In this role, I was thankful for the early experience at Holloman and working in a similar position with MSFC for Skylab. The CCB role provided an opportunity to learn about Orbiter Project Management from Aaron Cohen and the contractor personnel who were the pioneers of manned spacecraft design. My role on the CCB quickly brought me into conflict with several of the astronauts who preferred to work changes offline with Rockwell Downey. Aaron Cohen soon recognized the Rockwell change issues were bypassing his CCB and assigned me the action to review, and, if needed, to bring them to the CCB for formal disposition.

Aircraft and space systems are complex devices. In preparation for a flight in training and Korea, I carefully reviewed all "squawks" from previous flights, assessed fixes, if any, and made my flight decision. The same was true on the sign-off on the B-52 at Holloman. Aircraft sign-off is a risk judgment decision and is an acquired risk assessment skill. My experience interacting at the CCB on previous programs added to the critical knowledge base I needed for the pending Shuttle missions. Two days before a CCB, I received the listing of the CCB change topics; I distributed copies to the astronaut office and systems controllers; discussed the options; and established an approve/disapprove judgment on risk, cost, and schedule. This work complemented our mission rule development, and, by the time of the CCB meeting, I was well prepared to participate in the CCB, quickly earning critical Astronaut Office support for my work.

Shuttle Orbiter Project Manager Aaron Cohen

Cohen was born in 1931 of Russian-Jewish immigrant parents and graduated from Texas A&M in 1952. After his discharge from the Army, he worked for RCA and during that time, received patent credits for color television. He joined NASA and moved to Houston in 1962, where he managed the Apollo Command and Service Module (CSM) development before moving to manage the Shuttle Orbiter project. To provide visibility of my operations personnel, I would bring controllers to the CCB meeting to experience the program meeting decision process from the position of the engineering, Air Force, spacecraft, and contractor board members. On several occasions, I used them to speak at the meeting to justify the integrated MOD position, either for or against changes. At the meeting they observed Cohen, Faget, the Kennedy launch team, and Rockwell Downey run the tradeoffs, mission, and cost impact while summarizing our respective recommendations. This provided the controller with an understanding of the many shades of gray involved in even the simplest design change, i.e., every change had an impact on design, cost, schedule, and operability.

We were fortunate that North American Rockwell (NAR) was the prime contractor and we had previously worked together during Apollo with Cohen. Representing operations, I spent eight or more hours in the weekly CCB with all program elements. The meeting would convene at 10 AM and often stretch into the early evening. To keep the program moving forward, the challenge was to approve or to disapprove every design requirement, change, schedule, and assess costs at every meeting. In the early years, the issues were fundamental to design and, in the later years, originated from testing, integrated systems analysis, and flight operations.

One of the Air Force mission requirements established a design requirement for a highly unique mission profile. It required a launch, payload bay door opening, deployment of a device, door closing, and reentry to the Kennedy landing site in a single fractional orbit. This drove a high cross-range reentry capability and was a major design driver impacting heat protection provisions, payload bay operations, wing area increase, software, and others. This mission was never flown, but the impact on the shuttle was felt forever.

MOD established an operational requirement for an EVA capability, which was not in the original design specification. Designing for an EVA capability would require an airlock impacting weight, cost, schedule, and payload-bay volume. Cohen and the board fought this requirement for months. With strong Astronaut Office support, I finally won the issue by pointing out the complexity of payload bay door operations, uncertainties related to thermal soak, and the absolute need to get them closed and locked for reentry.

The CCB and Mission Rules process both addressed risk and utilized the engineering judgment of the best qualified personnel. Risk was addressed as one of several elements in system design to provide the highest reliability possible considering cost, schedule, complexity, and utility. The development of mission rules address impact of a unit or component failure as a part of a total system throughout all mission phases. The CCB and mission rule perspectives often merged during final system design when we would provide an integrated operational assessment when developing flight procedures. Arnie Aldrich had moved to lead the Orbiter avionics development and, when coupled with the MOD support, provided Cohen a critical, integrated resource for design study and major critical flight system assessments.

Approaching the final "design freeze," Mission Operations partnered with Cohen to keep the program moving forward by providing "Operational Workarounds," complex procedures solutions for recognized design deficiencies. The workarounds provided temporary solutions that the control teams and crews accepted. However, the design issue would be corrected later in the program. In August, the MOD, jointly with Aldrich and Kubicki in the Avionics and Engineering office, established provisions for in-flight maintenance. Within years, the concept of in-flight maintenance would become an accepted standard.

The synergism between the operations leadership that had moved into program office management roles and flight operations paid great dividends in the emerging Shuttle Program. The "task chemistry" developed with the program and engineering elements, like the SOCAR of Skylab, established technical relationships between the engineering, program office, and operational elements. It also provided the basis for the task and social chemistry, the "IT!" to successfully fly the space shuttle.

16

HOW TO LEAD YOUR LEADER

I have plenty of good people around me I can talk to and for me it's the man in the mirror that gives the answers if I have hard times or if I doubt.

—*Valtteri Bottas*

"IT!" does not sustain itself. Task chemistry is in continual need of renewal as the work changes, technology expands, and new capabilities emerge; all must be updated and reacquired. Character provides the foundation for the social chemistry component of "IT!" but to a great extent, it is generational, often shaped by the social mores of the times, events, and associations. When I joined NASA, the space program was new. Many of the applicants were military, already working in the aircraft industry or inspired by their professors in a time after Sputnik. I could quickly scan applications to review their postcollege work as well as search for time spent and skills developed.

In the two decades between forming Mission Operations in the 1960s to the 1980s and the post-Skylab period, there had been many changes in our nation. New recruits available for the shuttle program were recent college graduates. Few, if any, had served in the military or performed prior space-related work, so the principal gauge on their potential resided in their college transcripts and grade point average. Those applying for NASA employment came well-prepared in the sciences from major universities and were inspired by President Kennedy's lunar quest. Vietnam provided a character test between those who served their country and those with means who were deferred or moved to Canada. The civil rights movement and the movement for women's rights introduced more women and minorities into the workplace.

New employees displayed more confidence in themselves and their abilities, were gifted with new technology, had firmly established career goals, expected early leadership opportunity, and with economic growth, valued their free time. While they were not loners, it seemed they were looking for a personal, more independent lifestyle rather than a commitment to a working team. As a result, the new employees' social chemistry became more of a challenge to unit morale than in prior programs. The new employees' independence showed in the task assignments, work debriefings, oversight, and in personal development of software work aids that bypassed the MOD requirement for "software certification." The new personnel were technically brilliant, far above all prior new hires, but their independence showed in their relations with classmates and assigned organizations. While we spent time with the new hires briefing on our history, and the "culture" of Mission Operations, we quickly found that the one-on-one approach was not adequate to meet the challenge of the large influx of personnel for shuttle. This topic was addressed at a Directorate staff meeting and a discussion led to the decision to establish a "Boot Camp" for new employees.

Gordon Andrews, a young engineer in the Training Division, supported by key personnel from all MOD branches, established the "Boot Camp." It was designed to create confidence, teach passion for work, and build expectations. It included high-density classroom instruction; twelve-hour days; frequent testing; heavy mentoring; and emphasized organizational culture, values, teamwork, and technical excellence. The "Foundations of Mission Operations" document that I discussed in Chapter 11 was a common reference in defining expectations for the new employees, describing "IT!" and our performance expectations.

Andrews assigned me to teach a Boot Camp lesson on the Flight Director and team relations. I started with my criteria for Flight Director selection, focusing on the required abilities: leadership, team builder, judgment, ability to work under stress, good listener, and to be open-minded when possible. The Flight Director will also listen to what you say and the nuances in how you say it. Then I went to my favorite,

"How to lead your leader:"
As a controller, you must assume that he, the Flight Director is not very smart on your technical function,

he has only a headset, he has no data, and he has opinions based on his mission experience. You, as a controller, have discrete technical knowledge. You have data, training, and a team supporting you. Given time, you must convey your knowledge and thought process. Be patient and control the thought process. If you do not control it, it will control you. If you think the Flight Director is wrong, tell him. Do not leave your Flight Director guessing if you are not in sync with him. He has confidence in your ability, otherwise he would not have selected you."

The branch support to the Boot Camp was incredible because they knew they were training much-needed recruits for their organization and they got to work with the recruits before their assignments. Throughout the day, Boot Camp students arriving for class or finishing a tough series of tests would yell, "It's a great day to be in MOD!" When I first heard the shout, I considered it a sophomoric expression, but here was another example of "leading your leader."

"I grew to realize that the voicing of a cheer for the organization by the students established a collective unity of purpose."

Sometimes, students passing me in the hall would offer the "Great day to be in MOD!" salute. The approach was so successful that our contractors, as well as other civil service organizations at JSC, sent their personnel to our training program. After Boot Camp, the parent organizations continued their mission and the learn-by-doing, task-specific training. There was less than ten percent attrition during the training and the students gave high marks for the overall program.

By far the most dramatic example of "leading your leader" came in the way that progress in hiring forced us to see a whole new world of talent that was untapped in previous generations. Women were among the ranks of Mission Control from the beginning. Shirley Hunt, Mary Shep Burton, Cathy Osgood, and other members of the Mission Analysis Branch accompanied the mission teams to the Cape during Mercury, joining us for dinner, and time at the hotel pool. They supported software development, mission planning, and nominal and contingency mission planning. In early Gemini, the STG aircraft flight test experienced workforce merged with the introduction of those with "hands-on" experience growing up. Most came from small state colleges, were youths from the Indian reservations, or grew up on the hardscrabble ranches and farms of the plains. They were tough and competent and brought their passion and fire to the lunar journey.

It was the Shuttle program that introduced genuine diversity, providing the second transition from the all-male operations and instructor teams when young women initially sought membership and entered the ranks. Despite having a family with five daughters and a Mexican-American wife, I had never considered the emotional tax that came with the experience of being different from one's peers because of gender or race. This second transition introduced women with ambition, a thirst for learning, and the clear sense of direction that I saw in my daughters. Some women dropped out but those who stayed were pioneers who knew they had to have the right stuff, including tenacity to compete in what was an initially an all-male world. They did not, at least at first, want to be astronauts. They wanted to be controllers like the ones they saw on TV, part of the brain trusts that made it all happen. The women came from Virginia Tech, Purdue University, and other colleges, often with advanced degrees. Along the way, several had worked in the NASA Co-op Programs and were well experienced in an all-male culture. My "wingman," deputy Joe Roach, sensing the need to end the "men's only" culture, advised me that "IT!" must retune to a more professional and total team, one that welcomed women in our office areas. Within days, all inappropriate posters were removed from office walls to foster an inclusive work environment. Removing pictures was easy but changing attitudes took time. Still, everyone quickly learned that to sustain the organizational team chemistry and "IT!," MOD needed to institute a necessary and irreversible change.

Among the early women were Linda Patterson, Ann Accola, and Jenny Stein who each took different and extremely demanding career paths in mission operations. Looking back now, the successes were remarkable. Linda Patterson, born adjacent to the Langley Research Center, established her space

Shuttle GNC Linda Patterson

goal the day she saw the original seven astronauts in parades. She followed her father's path, graduating from Virginia Tech and believing "luck favors the prepared," she applied at every NASA center. The summer after graduation, a phone call from my Apollo 11 GUIDO, Steve Bales, anchored her dream. Linda was hired as a GNC (Guidance, Navigation, and Control), probably the most complex of the shuttle systems. After finding cardboard from the dumpster, she fabricated a full-size mockup of the crew panels to learn switch functions and came in on weekends to diagram the orbiter software flows. Linda reminded me of my first secretary, Sue Irwin, a rodeo rider. When males pushed the limit, both Sue and Linda played "hardball." Linda showed that the phrase "Tough and Competent" can apply to more than technical expertise. It can signify what it takes to change the all-male status to a new way of operating. Linda, in her retirement speech, spoke of her tenacity, "I was more determined to stay than the 'boys' were determined to force me out. Linda did double duty during the early Space Station program, supporting the initial work forming an organization for the mechanical structural systems used to attach new modules to the truss sections for the Freedom space station while in parallel preparing for the 1990 STS-35 shuttle mission. In other words, she was busy. Linda also helped to mentor the next generation hired to be tough, competent, and ready to pay it forward.

Jenny Howard, born August 8, 1955, came via a different path. Her parents were from the farms of Indiana and Kentucky. During the war, her father worked as a riveter on the Martin B-26 Marauder, a twin-engine bomber nicknamed the "Widowmaker" due to its high landing and takeoff speed. Jenny followed NASA from a distance during Mercury and Gemini. Entering Purdue University in 1973, she found there were few women in aerospace engineering. As a Co-op she worked at NASA Langley in the wind tunnel and later flew as a crewmember on the Boeing 737, evaluating automated landing systems. While working at Langley, she obtained her private pilot's license. When she graduated in 1978 and accepted a job at Langley, she was advised the dress code for female employees required a dress, hose, and closed-toe shoes.

In 1980, MOD was recruiting heavily for the shuttle program and Steve Bales contacted Jenny and recruited her for the trajectory team. One year before first launch, she began a career in Mission Control at the Booster Console. The booster position requires split-second irreversible decisions that are covered by twenty pages of mission rules and several charts. From the earliest days of the shuttle program, Jenny wrote the procedures, systems guides, console handbooks and certification guide. In her spare time, she played the viola with the Clear Lake Symphony and through night classes obtained a master's degree in Physical Science.

On July 29, 1985, Shuttle *Challenger* left the launchpad and was approaching six minutes in flight when the center engine shut down. The FIDO, Brian Perry, quickly assessed the Abort options and called for an Abort to Orbit (ATO) to get the shuttle into space at a lower altitude. Hughes, knowing a second engine shutdown might result in the loss of the shuttle and believing that engine shutdown was a result of a sensor failure, made a pure guts poker call to inhibit shutdown on the remaining two engines. *Challenger* reached a lower orbit successfully and finished an eight-day mission. The Mission Flight Directors selected Jenny Howard and Brian Perry to hang the mission patch in the control room at splashdown. In April 1992, I received the National Space Club, Astronautics Engineer of the Year at the annual Goddard Memorial Dinner. At the same dinner, Jenny Howard; and Flight Directions Cleon Lacefield and Brian Perry were also recognized for their quick action in 1985 which saved the mission by achieving orbit during the 51F launch.

Ann Accola was in seventh grade in the little town of Greeley, Colorado, when Sputnik was launched triggering an early interest in space. Under pressure to attend the State

Teachers College, because back then women either became teachers or nurses, she made the decision to attend Colorado State University whose mathematics curriculum included astronomy and computer science. Graduating in 1967, she joined JSC in a large class of over 300. Since few colleges had developed a space curriculum, she found herself unsatisfied. To get the education she wanted, she began attending Bill Tindall's Flight Techniques meetings and recognized that computer technology would give her a piece of the action, so she took a leave of absence to obtain a master's degree in computer science. She then returned to mission operations as an instructor in Space Shuttle Guidance, Navigation and Control. Anne Accola became the first female SimSup who tested the mettle of many Shuttle astronauts and controllers.

My point is not to say that these women were better than the men we would have staffed in such positions before, but to say that they were in all aspects *equal* to the men and that in recognizing this, we immediately doubled the pool of potential hires. Yet, like many of my generation, I was too "old-school" to see this on my own. As a leader, I needed to be "led."

1980

To staff the Mission Control teams for the first shuttle flight, I assigned the majority of my first-line supervisors because of limited controller training due to late deliveries of the Shuttle mission simulator software. The mission team was first class, reminiscent of the Apollo Program, from Headquarters through the Program offices and into Mission Control; the staffing came from the most experienced ever assembled for a mission.

For over three decades, all Flight Director selections except for two came from the ranks of the controllers. During Apollo, when I took my place at the Flight Directors Console, I was surrounded by nineteen individuals who were smarter than me. True experts in their areas of responsibility and I knew then that one day, one of those in the room, smarter and faster, would replace me. That day came on April 12, 1981. I was seated at the MOD console in Mission Control behind Flight Director Neil Hutchinson. That day, like when I left my Sabre on Taiwan, I was no longer a fighter pilot. Now I was no longer a Flight Director. I was seated on the top row as an observer, technically obsolete, but ready to answer the Flight Director's call for an opinion or support. As a leader, I had reached the point where it was time to gracefully take a seat in the shadows, giving way to a new generation to whom I would offer my opinion only when called upon. The fact that we could make this transition with ease made me proud.

Many organizations encounter difficulties in a changing of the guard but in a healthy organization, the sequence is seamless. That is as it should be, *the organization is the constant, not the personnel.* In this sense it is the organization that "leads the leader," not the other way around.

Hutchinson, a tall Oregonian with the visage of a leader, entered the ranks as a Flight Director during the Apollo 16 and 17 missions. He was a gifted operator with a math and physics background and was instrumental in the development of the Houston Mission Control telemetry and display system. He came through the ranks as Guidance Officer (Guido) and Guidance, Navigation and Control Systems (GNC) Engineer. In his oral history, he described his work developing the Houston Mission Control systems as a marvelous information exchange process that takes a tremendously complex set of information, does things to it, gets it to people who can make decisions on it, and then individual decisions are integrated and influence how a flight is going. Neil was "Flight" accountable to take any actions necessary for crew safety and mission success. I envied Neil, but my job as the MOD was to do the job that Kraft had done for me, which was to serve as the blocking back for Neil and the Mission Control team. There is a fine line in the relationship between the Flight Director and MOD and it must be studiously observed. The Flight Director is accountable for all the actions related to crew safety and mission success; during a mission there must be a single authority, the Flight Director.

The Shuttle Program first flight of STS-1 was a major milestone in the second era of manned spaceflight and of technology. I was present for all the first flights in the manned space program. After my years on the Shuttle change board, I had the confidence in the design, testing, and in my operations team that came from knowing leaders who always placed crew safety first in the design decision process. The shuttle first flight demonstrated the power of the computer-based technology available to engineers to design complex, highly integrated and reliable systems for flight. Knowing the people, I never doubted that when we launched the STS-1 crew they would return. STS-1 was a "real flying machine," half-spacecraft and half-aircraft, flown by computer, reusable, and landed as an airplane. As MOD for the first Shuttle mission, one of my duties was to lead a small team to assess any thermal protection tiles lost or damaged during launch using images from several sources.

There are four communications plug-in locations on the MOD console. I normally had the JSC "brass" at my console during critical mission periods. For the STS-1 landing, Max Faget, the pioneer spacecraft designer of the early programs, and Chris Kraft, the leader and operations boss who wrote the book on manned space operations, were seated with me. Faget followed the displays intently during reentry and, as the Shuttle emerged into the subsonic flight regime, Max could not keep his mouth closed nor could he contain his enthusiasm. Kraft nearly gagged him to keep him quiet. Max, like Bill Tindall who sat with me for the lunar landing, was a giant in the annals of space and it was my privilege to have him sit with us that day.

The Shuttle Flight Directors and controllers who were tested and grew to maturity in Skylab operated the most complex space system in history. The post-mission reports indicated the need for the time to acquire and train engineers, controllers, instructors, to achieve skill levels greater than that ever achieved during the Apollo program.

The shuttle was a technological marvel, and its first mission involved a level of risk equivalent or greater than John Glenn's historic mission.

***

The NASA Administrator is nominated by the President and confirmed by the Senate Committee on Commerce, Science, and Transportation. Through the first two decades of space leadership at the top-level in NASA Headquarters provided a stable and predictable base for the manned space programs. To a great extent, this occurred because the structure of operations, program offices, and NASA Headquarters personnel all grew together, becoming a team who were familiar with each other and comfortable with assigned duties, roles, and missions.

With a decline in defense funding and a narrowing of commercial markets in the post Apollo period, aerospace competition for each project became more intense. With the aircraft and space industry consolidation, NASA could no longer depend on industry to provide leadership like Dale Myers, John Yardley, or Mike Malkin to help us start programs. Several leaders who arrived in the Shuttle-era arrived unaware of space program history, program roles and missions, and while experienced in corporate functions and finance, were unsuited to lead a massive operations organization. I believe the sea change—the transformation in NASA leadership—began early in the Shuttle program. The start of the flight phase of the Shuttle Program was very turbulent. From 1980 through 1982, the leadership in Washington changed three times from Robert Frosch to Alan Lovelace to James Beggs. On July 10, 1981, midway in the seven-month period between the first two Space Shuttle launches, James Beggs was installed as the NASA Administrator. Beggs, a Naval Academy graduate in 1947, had served in the Navy until 1954. After graduating in 1955 with a master's in business from Harvard, he served in several corporate and government positions prior to entering NASA as Administrator at the beginning of the Reagan administration. I personally briefed Beggs on several occasions and quickly formed the opinion he was a good man, who held a general belief that NASA's leadership suffered from poor administration, lackluster business acumen, and did not possess the ability to achieve the shuttle program's advertised goals of a high flight rate and low operations costs. Beggs had a limited knowledge of space operations and the responsibility of the Mission Control teams in controlling risk through planning, training, and during mission operations. I was surprised by a major leadership change so early in the in-flight test phase of the Shuttle Program.

The Headquarters STS-2 Flight Readiness Review was conducted in Washington on October 22-23, 1981. During the review, I briefed Beggs on the status of all mission support elements, crew and controller training, the flight plan, mission rules, and closeout of the STS-1 debriefing actions. I concluded my briefing with "Flight Operations is Go for the STS-2 launch." The second Shuttle mission with the crew of Dick Truly and Joe Engle was launched on November 12, 1981. I was serving as the MOD and JSC Center Director Chris Kraft was sitting with me at the console at liftoff. On the second orbit, two Shuttle Orbiter fuel cell discrepancies were noted. The controllers major concern was a high pH, (basic) message telemetered from Fuel Cell 1, although performance at that time appeared normal. Subsequently, the fuel cell performance decreased by 10% confirming some internal problem. The concern among the flight control team was a possible breakthrough of the fuel-cell membrane. The fuel cell was shut down and the oxygen and hydrogen valves were closed, thus saving the system, and avoiding a potentially explosive mixing of reactants. The problems of Apollo 13 were still fresh in all the controllers' minds.

Several hours' discussion ensued between the controllers and Rockwell engineers relative to fuel cell similarities with the Apollo systems. The mission rules indicated that with loss of a fuel cell, the mission would revert to a minimum duration with reentry scheduled for the next day's planned landing site. Sitting at the adjacent Headquarters console was the NASA Deputy Director, Hans Mark, who was taking detailed notes of events and times. His occasional questions indicated a grasp of what was going on and the flight teams' concerns.

Neil Hutchinson, the Flight Director, turned to me saying, "I have talked to EECOM, and SPAN and by the mission rules I am going to a minimum mission. I would like to brief the crew prior to sleep." This shortened the planned five-day mission to two days. I nodded concurrence to Neil's decision and the mission team initiated minimum mission planning.

Administrator Beggs, in Washington, had been monitoring the discussions and established a phone line conversation with Kraft and me from his home. The lengthy discussion was a distraction to Hutchinson, so we moved to the viewing room to continue our talk. I reminded the Administrator that this was only our second flight, we had limited flight experience on the Shuttle fuel cells, and there was a lack of maturity in our two fuel cell reentry procedures. Beggs diverted the conversation to discuss fault trees and failure history, of which we had none, and the impact to the image of the program. The discussion quickly got down to our insisting that, "Dammit, we've got the best people in the world working the problem. Trust us!"

Shift change time and the crew sleep period was fast approaching. I again went to the control room to tell Hutchinson that we concurred on his minimum mission and to inform the crew of the decision. Kraft impatiently continued talking to Beggs, finally advising him that it was decision time, and concluded the conversation with, "I support the Flight Director's minimum mission duration decision followed by reentry." At Hutchinson's request, I then briefed the Mission

Management meeting on the mission rules and the content of the minimum mission.

Returning to my console, I watched Hans Mark review his notes from the Management Team meeting after which we started talking. In many ways, Hans was reminiscent of the great NASA leaders of the past and had the personal and inspirational qualities of my long-ago mentor, Harry Carroll.

Hans Mark, Deputy Administrator, and Dan Germany

Mark and I were close in age. He was born in 1929 in Vienna, Austria, and escaped Nazi Germany in 1938, immigrating to the United States. His education took him to the University of California and a PhD in physics from MIT followed by service at NASA Ames, National Reconnaissance Office, Secretary of the Air Force, and NASA. Upon leaving NASA in 1984, Mark served as Chancellor of the University of Texas system until 1992 before moving to a research position at the Pentagon. In 2001, he returned to the University of

Texas where he held the John J. McKetta Centennial Energy Chair in Engineering.

Hans was personable, thoughtful, and possessed an enormous energy and passion for space. I always looked forward to meeting him again during missions or while at Headquarters for the readiness reviews. His book, *An Anxious Peace*, at seven hundred pages, provides a graphic history of the Cold War period from one who lived it. Mark quickly became my "operations broker" with Administrator Beggs, saving me "face time" on trips to Washington to explain the Mission Control rules and decision process. We became good friends and communicated verbally and in writing for the next two decades.

Unlike prior administrators, Beggs spent much time questioning operations and the mission preparation process. He was learning about Mission Control, the team, and the functional responsibilities on a mission by mission basis. I believe he knew that the advertised flight rate and the cost efficiencies would never be achieved, but it was his job to press operations to find ways to lower costs. I doubt that he ever understood that the flight costs involved the issue of accommodating the large span of vastly different payloads. The standard satellite deployments from the payload bay were a piece of cake, however the majority of payloads required tailoring of flights to optimize the science requirements requiring significant mission planning and training. Years later, in my final briefing to Beggs, I spoke of the planned contractor consolidation and the predicted cost saving. I finally obtained his total interest; we were on a subject that he mastered.

Throughout 1982, Kraft repeatedly told me it was time for me to move to the Program Office. Initially, I just thought that it was an off-the-cuff comment. In the following months, he again mentioned moving to the shuttle program forcing me to address his question. I think that he wanted me in the Shuttle Program Office before he retired. I was not afraid of change, the question was, would I make as good a project manager as Kleinknecht, Thompson, or Cohen? I admired every Program Manager I had worked with and respected their work. Program leadership, however, involved the management and direction of the contractor team in developing the systems capability, the rockets, spacecraft, and support systems to accomplish a defined project.

My life had centered on operations since my fighter pilot days. I had clearly worked with and participated in many aspects of program management during my CCB work with Cohen and Kleinknecht. I knew I would have to become a different person if I selected a direction to manage programs. My life was operations, and my skills were best utilized working with people, leading teams, communicating energy, and challenge. That was when I was at my best.

In August 1982, at the conclusion of the Orbital Flight Test phase of the Shuttle Program, Chris Kraft retired. Chris sent me a letter and came down to my office to advise me of his retirement decision two weeks prior to the announcement. He indicated concern for the shuttle program after he left and again indicated that I should move to the Shuttle Project Office. He never provided a reason for his retirement, but his discussion indicated he believed that pending changes in contractor assignments would impact the hands-on knowledge at the Cape rendering mission work in the coming years more difficult. He concluded wishing that whatever happened, I would, "*Stay strong in my position.*" The "stay strong" words were strange, but in coming years I would soon see them as prophetic.

I was shocked to see the JSC leadership change before the program had reached a stable point. We were still flying two-man crews in shuttles equipped with ejection seats and soon had to step up to a significant flight rate increase. He did not mention a replacement, and I wondered if he had been offered the choice to select his successor. Frankly, I thought only he had the knowledge to select a replacement. To me there were only two people capable of the job, Glenn Lunney and Aaron Cohen. They had the big picture, but neither had the full range of personal and team management abilities of Kraft and would have to grow into the Center Director's job. In the earlier programs, NASA could often draft experienced aerospace leaders like John Yardley, George Jeffs, Dale Myers, and Mike Malkin. However, the aerospace industry no longer had the breadth in leadership, and I doubted that a search would find a replacement capable of stepping into the job this early in the Shuttle program.

JSC was fortunate, however, that when Kraft departed, he left a strong but thin NASA Center leadership capable of operating the shuttle system. I wondered how long that thin line of leaders would last? In the early years, Chris carried the nickname "teacher." Chris' loss would be extremely difficult for all those whom Chris had mentored, tutored, and had taught to assume great responsibility. I thought back to the days we had shared an apartment at the Cape. He developed individuals who understood the nature of risk and were willing to step forward and say, "I can do anything you ask, even if what I think is impossible . . . and when it is no longer safe, I will say 'stop.'" Where would we find the leaders to step into his shoes? For over two decades, I watched Kraft grow, accepting great responsibility and providing the role model for those willing to accept accountability for their risk judgment. I doubted there would ever be another like him. The Shuttle was the last hurrah of Kraft's generation.

17

OPERATIONS

I am ready to work, I am ready for this job, and I am ready for this challenge.

—*Stuart Pearce*

Leadership is a journey that lasts a lifetime. There are many forks in the road, many decisions to be made, and many times you will need to retrace a path and begin again.

Throughout this journey, you will often be faced with the challenge to remain true to your beliefs and, in the process, you will struggle through difficult periods. At times, you may question your path, your convictions, your approach, and during these times, you must rely on your moral compass to see you through to your destination.

On October 14, 1981, a year before Kraft's departure, I briefed the HQ Management Council that the higher flight rate demanded a close working relationship between the program office and operations in defining mission content and schedule. Operations needed to be a partner in the payload manifesting process and with the program, we must jointly provide control over the requirements and late customer payload deliveries.

Gerry Griffin had been off my radar since he left flight control in 1973 and transferred to NASA HQ as the Associate Administrator for Legislative Affairs. He made the rounds of Dryden, Cape Kennedy, and then, over a nine-year span, returned to Headquarters. In the summer of 1982, rumors started that he was returning to JSC as the Center Director. If true, it was time to celebrate the continuation of continuity in operational leadership at the center.

Griffin was assigned as JSC director in August 1982 and quickly filled the gaps formed by Kraft's departure. In March 1983, he restructured the Mission Operations Directorate into two functional elements: one supporting the astronauts and the other with overall mission operations, facilities and software responsibilities. Abbey was assigned as the Director of Flight Crew Operations covering the Astronaut Office and the Aircraft Operations Division. I was assigned responsibility for operations, training, facilities and all software including the space shuttle. This reorganization formalized the way George and I had been running the directorate for the previous seven years. I now had the clear responsibility for Shuttle Manned Spaceflight Operations, and, with it, the accountability. Griffin's reorganization further anchored the Mission Operations Director role in Mission Control.

This reorganization was the first of two moves in Griffin's overall plan to consolidate and streamline all Shuttle operations under a single contractor entity. Except for the Program Offices, the basic structure of the JSC contractors supporting operations had been virtually unchanged since the completion of the Apollo program. This resulted in a proliferation of contracts and contractor organizations. While still trying to process the total job, particularly budgets, funding, and resource management, I was advised by Griffin that he planned to consolidate JSC contractor organizations. The proposed Space Transportation System Operations Contract (STSOC) merged sixteen contracts in MOD into a prime contractor with two subcontractors, involving about 5,600 personnel. The contract transition was planned to start on January 1, 1986, the year the planned flight rate was scheduled to increase to fifteen flights per year.

With the shift of the Shuttle Program to the operational phase after the STS-4 mission, the program office delivered a torrent of payload requirements and changes to the mission manifests. This was not a surprise since the "payload user (customer)" phase had been delayed until the flight test phase was completed. The near-term missions, included the first EVA, multiple satellite deployments, first night landing, and deployment of a tracking and data relay satellite (TDRS) that required real-time interaction with the DOD facility in California. My staff and I had our hands full to say the least.

With over two decades of spaceflight experience I was now provided the opportunity to build and lead a high performance flight test organization tailored to the shuttle high flight rate, user-oriented program. Through the years I have operated in two organizational modes. The "home" office organization is the backbone for personnel acquisition. It is structured functionally, systems, training, planning, etc., and the individual skill building is accomplished "learning by doing," developing all materials needed for flight. Flight Directors, when assigned to a mission, operate across the functional organization to provide the long lead time materials to support early mission policy decisions. Mission teams are formed about three months before launch and work as a cross-functional team, laterally across disciplines to provide flexibility, mobility and integrated solutions and develop IT! the mission task and social relationships and team unity for flight operations.

One of my initial actions in preparing for Shuttle operations was establishing an executive secretary for operations in Mission Control. When I was assigned as the Assistant Director of Mission Operations, I recruited and selected Jan Pacek as my executive secretary in 1975. Her knowledge and hands-on technical skills grew rapidly in preparing post-mission reports, actions from debriefings and the Orbiter Change board. She was a good teacher who established the administrative support standards for mission operations. During mission preparation and operations in Mission Control, she supported the entire NASA leadership staff and adapted quickly to the needs of any situation she encountered. Pacek was a Texan, born in Odessa. Her father was an oil field roughneck and farmer. Many of my controllers came from the small communities, farms, and ranches Texas and Oklahoma. Pacek was no exception, her hands-on experience came from working in the watermelon and cantaloupe fields. She began at JSC in 1969 and her "quick study" abilities fit well into the technical aspects of space systems and operations. Many times, while I was participating in meetings, Jan would maintain my mission log, crisply summarize technical issues and actions. She functioned well, often executing actions, relaying information, and preparing correspondence for top NASA management. She quickly became a critical element in every aspect of directing my organization. Fifty years later, Pacek continues her support, typing, providing research, answering correspondence, and providing an "independent" assessment of my work.

Jan Pacek at the MOD console in Mission Control

The top-level NASA Headquarters leadership provided a stable base for mission operations through the first two decades of the manned space programs. This, to a great extent, occurred because the structure of operations, program offices, and NASA Headquarters all grew together. Retirements, however, reduced leadership depth, operations expertise, and program continuity in many areas. Replacement personnel lacked the necessary operational abilities; change was in the wind.

CHALLENGE

18

GO AT THROTTLE UP!

History is a vast early warning system.
—Norman Cousins

During the winter months growing up in Toledo, the kids on Berkeley Drive would go down to the frozen Ottawa River to play hockey. There were plenty of broken tree branches and we would start a fire in a rusty fifty-five-gallon drum, dust snow off the ice with our brooms, and form teams. I have been a lifelong hockey fan and recently purchased Mark Messier's book, *No One Wins Alone.* It is an excellent story of leadership and what it takes in terms of individual performance and team commitment to become a "dynasty" in the sport of hockey. Playing as an Edmonton Oiler, Messier lifted the Stanley Cup aloft on four occasions, and in 1994, playing at Madison Square Garden for the New York Rangers, he led his team to a game-seven win and the Stanley Cup championship—the first for the Rangers in fifty-four years. It was a tough series, but the players went beyond skating for each other. They skated for the fans, the management, for New York, for the franchise, and the Rangers. It was a display of "IT!" writ large.

Messier's description of the character, grit, and determination necessary to win the Cup in 1994 is a compelling read. But it is also a statement of how an organization, even one as great as the New York Rangers, can come up short. After winning the Cup in the 1994 game, a change in Rangers' management, combined with injuries and player trades, resulted in a team running on fumes. They came close but they were never able to achieve the level as a champion again. The Rangers had lost "IT!"

Over the decades, the Flight Control Division emerged as a dynasty and, during Gemini, it was recognized as the core element of NASA manned space operations. The division retained this position through Apollo and then saving Skylab in 1973. However, at NASA, as with the Rangers, management was changing. The shuttle was the last major program effort of the Mercury generation and most of the founding members: Gilruth, Faget, Kraft, Slayton, Thompson, and others had retired. The space industry had also gone through several consolidations and the likes of John Yardley, Dale Myers, and Tom Kelly had departed. With Deputy Administrator Hans Mark's departure in September 1984, I lost my last trusted direct contact at NASA Headquarters. I had an uneasy feeling that the Headquarters leadership paradigm shifted away from operations safety to a high flight rate, that focused on the customer's needs, be that customer the U.S. government, academia, or a commercial entity.

My directorate provided the planning, training, and manned space operations and is probably best described as a mission factory with preparation for many missions running in parallel. To stay in touch with the factory floor, I would often sit in on mission rule sessions or in the viewing room during training and listen to the debriefings to assess team performance. The 61-C Shuttle mission, which included Florida Congressman, future U.S. Senator, and NASA Administrator Bill Nelson as a crew member, slipped its scheduled launch nine times during the twenty-five-day period over the Christmas holidays. If we had launched on schedule, we would have flown ten missions in 1985. That was already ambitious. Now, with no slack in the schedule and limited resources, fifteen flights were manifested for 1986. The only way to succeed was to depend on "IT!"

The management at NASA Headquarters works at a distance from those in operations and can lack the feel for the impact of the consequences of their decisions. With hundreds of engineers and technicians changing plans, rearranging family schedules, missing time off for weekends and holidays only to encounter yet another change, I began to wonder whether we could do it all. It was easy to change the schedule. Yet the astronauts, the facility technicians, and the controllers were the ones regrouping to make those changes happen. "IT!" and the NASA character, grit, and determination were keeping us going, but like the Rangers, were we getting perilously close to running on fumes? Kraft's words, *To stay strong in my position,* were now frequently coming to mind.

You will remember my first job with Harry Carroll was assessing telemetry data. Harry taught me that by analyzing the data on flight tests, I would soon know more about what happened than the pilot did. I learned that lesson well. Ever since then, I have been a believer in using time and task metrics at the division, branch, section, and individual controller level to assess performance and identify any deviations from the expected profile. As Division chief, I annually produced a "State of the Union" memorandum that was distributed within JSC addressing the prior year's mission operations performance in the same way that I looked at the McDonnell flight test data. To my knowledge, my division was the only organization at JSC, maybe even NASA that collected daily performance data. My annual State of the Union provided metric data on every aspect of the MOD mission production and was normally produced in early January. However, with the pending transition to the Space Transportation System Operations Consolidation (STSOC) contract and my concerns with the 1986 manifest, I provided the forecast on July 2, 1985.

The State of the Union memo was addressed to Cliff Charlesworth, Kraft's Director of Space Operations. The 1985 data clearly indicated that the year's flight rate was accomplished by the reallocation of personnel, flight planning, procedures, and simulator resources from the 1986 missions to support the 1985 missions. The 1986 crews, Flight Directors, and teams were behind the power curve. They would work with incomplete and late flight data files, reduced simulator time, and compressed mission support schedules. To fly the manifested 1986 program, a major revision of the manifest and schedule was required and I would need strong support from the Astronaut Office to obtain Charlesworth's concurrence to carry the issue to the Center Director and Program Office. I attached a note to my "State of the Union" message addressed to Charlesworth that ended with the words, "No matter how well we manage, we are approaching the end and we will see it in the next twelve months. Spaceflight is utterly unforgiving of error." Six months passed and there was no action to address the flight rate and manifest issues. I found myself remembering my Apollo 1 warning to remain "Tough and Competent" as a bulwark against error. Did we need to relearn that lesson?

1986

The STS-61C launch slip from December into January was a key topic of my January 2 telecon opening new year 1986. On January 12, after many delays, 61C finally lifted off. The morning telecons normally ran about half an hour and were of a "negative reporting" nature. In other words, there were to be no status reports, just updates on issues that were truly a problem. Two items dominated early January, the ripple effect of the launch schedule changes on the missions in production and the high-risk Galileo/Ulysses missions. The Centaur stage, used for both missions, was not designed as a manned system and did not possess the critical redundancy utilized for manned systems design. I also believed that the changing management at JSC and at NASA Headquarters was a further indicator of pending troubles. Because of late software deliveries, the 61F/G Centaur simulation schedule was currently behind by six weeks and the crew-integrated training flow would be limited to an unacceptable thirty hours each for integrated and standalone training.

The Ulysses and Galileo missions required two launches in a five-day period beginning on May 15th. The missions carried nuclear generators and utilized the thin-skinned liquid Hydrogen/Oxygen Centaur stage which required unique handling, shuttle structural modifications, and complex propellant-dump systems. The two missions were often referred to as the "Death Star" flights because their risk encompassed far more than the obvious dangers associated with a Shuttle launch. Safety issues were a topic at every mission review. In my mind, the issues in January 1986 were reminiscent of Walt Williams's comment on the failure of the first Mercury Atlas launch. "We launched into an on-time failure." I do not know how many of the Mercury team were still working on the Shuttle and would remember Williams's debriefing words, but I remembered those words every time I gave a mission "Go."

There were other factors complicating the environment. January 1, 1986, was the effective date of the STSOC transition and most of the mechanics and paperwork of the transition was on schedule. I knew that there would be some major glitches, but I was confident in my directorate leadership and their ability to address and exercise risk judgement when needed. The next six months would assess the MOD team's ability to sustain the "IT!" During the transition, the workforce was faced with the most complex mission sequence of the Shuttle flight program to date. The civil servant component was ready, but the weakness, if any, would show in the contractor workforce. The inability to transfer the flight-design team from the McDonnell contract left STSOC with a thirty percent experienced trajectory workforce with some major skill deficiencies. The Cape team was also facing the challenge to support the first launch of a DOD shuttle mission from Vandenberg, which I would be supporting in Mission Control with an Air Force "Blue" team I had been developing for several years.

I was in Mission Control with Tommy Holloway on January 12th when STS-61C with Rep. Bill Nelson lifted off on the seventh launch attempt. "Grey Flight," Gary Coen, was leading Mission Control. Gary had come through the ranks in Mission Control to become the 15th NASA Flight Director. He studied automotive engineering at the General Motors Institute (now Kettering University) and when joining NASA, was assigned as Gemini Guidance Navigation and Control for Gemini 4. When Gary served with Griffin and Aldrich during Gemini, they formed the strongest team we ever had in the GNC position. Gary supported me as GNC on my first mission as Flight Director. He was a heavy smoker who always seemed in need of a haircut, looking like a fisherman getting back from a day on Galveston Bay. He was a superb controller and Flight Director who seemed oblivious to the world outside Mission Control.

With the advent of the STSOC contract, I transitioned the MOD Mission Control console function to the Flight Directors Office and had been alternating mission support between Tommy Holloway and Don Puddy. Three launch-day slips moved the launch of 51L from January 22 to 28. A widespread Arctic air outbreak in the Eastern U.S. on launch day resulted in an overnight low temperature of twenty-six degrees. Leaving my briefcase in the office, I walked around the perimeter of the duck pond to Mission Control, arriving about 6:30 a.m. to join Puddy at the console. I arrived as the ice teams were providing a stark description of the ice hanging on the fixed services structure, mobile launch platform, and pad apron. Believing there would be a launch delay, I briefly left Mission Control for a walkthrough of a new portable building that had been constructed to collocate my shuttle flight planning and procedures team.

Returning to Mission Control, I noted that the *Challenger* pad video was grim. As I listened to another report from the ice team, I had confidence that the MSFC team and especially George Hardy would make a launch "No Go." One hour prior to launch, I was joined by my secretary Jan Pacek, Cliff Charlesworth, and Gerry Griffin.

Liftoff from Pad B occurred Tuesday, January 28, 1986, at 10:38 a.m. CST. Launch appeared normal in Mission Control. The launch phase through SRB separation is the toughest part of every shuttle flight. There is no crew escape during

SRB burn. At engine throttle up, I relaxed, as soon we would be off the SRB's. As the shuttle passed maximum dynamic pressure (Max Q), they received the "Go for Throttle Up" from the CapCom. Seconds later, at an altitude of 46,000 feet and seventy-two seconds into the flight, there was a huge fireball. One second later, the shuttle broke into several large pieces. Range safety confirmed vehicle breakup and soon thereafter, the Flight Dynamics Officer reported tracking several pieces.

The Challenger Shuttle explosion

For what felt an eternity, I watched the video picture of the *Challenger* loss, then stepped down from the MOD console and went to advise the procedures officer to initiate the mission contingency checklists. I will never forget the look on the faces of Flight Directors Jay Greene, Lee Briscoe, and CapCom Dick Covey. They had lost their crew, all of them very close friends. In the years that followed, I knew that day changed Jay Green forever. Jay had been my FIDO during the first lunar landing.

The hours that followed were reminiscent of Friday, January 27, 1967, when we lost the Apollo 1 crew, only this time the whole nation was going through the awful pain of loss and guilt. The crew included Christa McAuliffe, a high school social studies teacher from Concord, New Hampshire. The effect of her loss on schools and educators was conveyed in a condolence letter sent to mission operations, January 30, 1986 from the Alhambra School District in California. They summarized the grief of a nation, "While our flags were being lowered, our hearts were grieving at this tremendous setback."

For the new generation of controllers, the "shuttle generation," there was no Singing Wheel to let alcohol dim the pain nor did they have the extremely close camaraderie among them to soften the blow. Many of them had never been present for a failure when the crew died. They were just left to watch the endless playbacks of launch video and the breakup, then go to their offices or home with pain in their gut and ask the questions. *What happened and why?* Prior to launch, Ellison Onizuka often provided pineapples that were delivered to the cafeteria for a pineapple upside-down cake celebrating the mission start. We had the cake sent to the local boy's home. Astronaut Ron McNair played the saxophone in MOD's "Big Band" at my daughter Lucy's wedding. He and the other crew members were now gone. The social aspects of "IT!" made the crew loss up close and very, very personal. Within hours, the media and television begin dissecting the actions of the organizations and personnel involved. For many days thereafter, the video of the explosion and death of our crew would be replayed endlessly.

After a disaster of this magnitude, it is important to be proactive and provide a perspective to counter the normally negative initial report. During Apollo 13, initial reports addressed crew loss, but within hours we were listing options and, within twenty-four hours, were describing our plan to save the crew. This offered no such opportunity. Coping with the trauma and the shock of the death of our *Challenger* crew that only hours before had walked from the crew quarters, entered the transit van, and then were strapped into the shuttle was in many ways traumatic.

The risks are debated in the mission rule sessions and during integrated simulations and the readiness reviews. We know of the risk, but it always seemed remote. We accepted it as our job.

Sustaining "IT!" after a crew loss or major failure starts with leadership at the top. The first step for me and my leadership team was to clearly recognize that the *Challenger* crew

died for a cause; they were the explorers, and we were their guides. We would share the grief as a team, we would find the cause, and to the best of our abilities, we would assure it never happens again. We would be "Tough and Competent," and, like our early Apollo predecessors, we would not fail in our mission duties.

Yet in the back of my mind, there was the awful recognition that we had been here before. I and the others had sworn that by being "Tough" and "Competent" we would ensure that it never happened again and that, despite our resolve, it had happened.

This was the first crew ever lost in flight and this time, I did not wait three days for NASA management to provide direction as I did after Apollo 1. On launch day, I had an organization of over 5,600 personnel who were in the process of changing jobs, and leaders who were just reforming their organizations. A recovery plan and a clear sense of direction was needed quickly. Returning to my office, I grasped a red Sharpie and began writing. (I didn't remember that it was red until in writing this book when I returned to look at that day's notes.) I had the first major crisis of the new MOD organization. "IT!" was critical. I needed to fire up the new generation of controllers and get them engaged and moving. Within a half hour, I scrawled a five-page meeting agenda that began with the words "We will fly again!"

We are leaders and we must lead. There will be confusion and conflicting actions and strong criticism. Our people have not been through this before. It will require leadership. We must be the stable element.

Four hours after the loss of *Challenger*, with the civil service and contractor staff, I began what may have been the most important meeting of my life. I reiterated "We Will Fly Again." The STSOC transition had been initiated January 1, twenty-six days before the accident. My first actions established a clear direction for the STSOC contract, rapid assessment of the anticipated contract budget impacts, and a plan for the contractors who had not completed transition. I established a war room to address the personnel and contract issues resulting from the accident. Within days, a plan emerged. Several of the manifested missions were in production in MOD and even though we were not going to fly, continuing the flight data file production would provide the opportunity for training the new contractor teams. Prior to the accident, I had been using most of my supervisory staff in key Mission Control positions, but now simulators and computer resources would be available for training. A major thrust for the new STSOC team was to achieve certification of 100 percent of all new personnel for mission support during 1986. We elected to continue the work in the trajectory design cycle of the existing missions to provide training of the new personnel in the critical flight design skills. Finally, the Flight Directors began a formal review of the "All Flight

Rules" document that verified the rationale for each of the rules. I participated in a weeklong review of the rules and associated rationale with the Flight Directors. This was my most enjoyable task in a challenging time.

The STSOC contractor reorganization did not address a span of control issues in my Engineering and Maintenance Division (EMD). The division had the responsibility for design, development, operations, and sustaining engineering for the full range of hardware and software for mission operations. This included the pre-mission trajectory, flight planning and the Mission Control room and training facilities. The Orbiter AP-101 computer flight software and the related day of launch reconfiguration was the most critical new work element. In the MOD tradition, my last action was to schedule a date to hang the 51-L crew patch in Mission Control.

A widely distributed "memo for record" by John Young, the chief of the Astronaut Office, provided a real boost to "IT!" John recommended that my flight controllers should be assigned to directly support the Program Office Systems accident assessment. This recognition was a real "atta boy" and, as a result, almost half of my systems controllers were assigned directly in support of the reassessment of the orbiter systems, procedures, failure modes, and operational mission requirements. The chief of the Flight Directors Office, Tommy Holloway, was assigned as the FOD lead to support the 51-L investigation.

At the time of the accident, NASA Administrator James Beggs was on a leave of absence dealing with a federal indictment for an event that occurred prior to entering NASA. He was eventually completely exonerated. Acting in Beggs's place was William Graham, a physicist with no knowledge and experience in space, whose appointment had been vigorously challenged by top NASA leadership. In an overnight phone call with the White House the day of the accident, Graham agreed to establishing a Presidential Commission on the accident, one that would be led by former Nixon-era Secretary of State William P. Rogers. This was in marked contrast to the process used to address the Apollo 1 fire.

Within a few days, the accident video investigation centered on a failure of a Solid Rocket Booster (SRB) as the accident cause. Other than periodic status reports, my personal directorate workload precluded significant involvement in the investigation process beyond establishing the flight rate assessment.

Astronaut Richard Truly became the Associate Administrator of Space Flight on February 20, 1986. A month later at the JSC staff meeting in March, he addressed the strategy for returning the Space Shuttle to Flight (RTF). His speech provided the needed message that flying in space was a bold business and he intended to correct any mistakes and return to flying soon. Truly, at that time, did not know the Commission report would quickly change from finding and addressing the technical issues to broadly addressing the NASA organization and the decision-making procedures. The final report addressed the technical aspects of the solid-rocket history well but failed to clearly address responsibility. Subsequently, the report evolved into a discussion of the NASA organizational culture and recommended that the solution to install a "safety culture" was to place more astronauts in management roles. I considered this a placebo since flight safety was always a principal consideration of the Program Level II and Project Level III change-board activity and the JSC Safety Office.

I was a board member and reviewed every change with the mission controllers and provided distribution of all change material to the Astronaut Office. To my knowledge, the issues with the SRB field joint were never addressed at the Level II or Level III boards. This is not an indictment of the process, it is an indictment of the presenter and the organization they represent. The astronaut office and Flight Operations can only work and take actions with the technical engineering data provided. Time is money, so NASA, under Truly's direction, moved out briskly to address the technical issues well ahead of the Rogers Commission to keep the people, program, and budget on track.

Beggs resigned on May 12, and Truly got a new boss, the former administrator, James Fletcher. When the Rogers Commission Report was released in June, it pointed to a design flaw and described the event as "an accident rooted in history." It was from this that I learned the concerns with the motor case joint and seals had been first documented in 1977, six years before *Challenger*, but had never been addressed at the flight-readiness reviews. After assessing the management communications on the night before the *Challenger* launch, William Rogers advised the press that he believed the decision process itself was flawed. This indictment of Shuttle management, while not limited to the mission launch decision, wreaked havoc within NASA and provided the momentum behind the Commission's recommendations on the need for widespread organizational change. The post-*Challenger* NASA restructure became like the game of "musical chairs," changing program management roles and inserting astronauts in the NASA leadership chain in key locations to "solve" communications and provide a "safety emphasis."

As Director of Mission Operations, I was involved with upper management principally during the major reviews and the formal annual budgeting process. The changes that directly affected MOD work involved JSC Director Griffin's retirement. Robert Goetz replaced Griffin for a short period before being replaced by Jesse Moore the day of the *Challenger* accident. By October, Moore was back working in Washington and was replaced by Aaron Cohen as Center Director. Mindful of the disruptive atmosphere, Aaron Cohen described this period as "A Time Worse than Chaos. . . . For a period of time, JSC had no spokesman and NASA had no leadership."[14] Lunney retired from the shuttle program and was replaced by Aldrich, who was relocated to NASA Headquarters as the Manager of the National Space Transportation System (NSTS). Dick Kohrs, his deputy, remained at JSC and, fortunately, we had a solid working relationship.

With limited top-level leadership, lower-level leaders, to the best of their ability, were running around, finding and trying to plug the holes. While MOD survived through "IT!" during this difficult period, NASA's credibility was tarnished. Confidence had been shaken, morale was low, and leadership was completely displaced. During the launch delay we trained contractor personnel in the trajectory skills in time to return to flight, certified the new STSOC personnel, and vastly improved the flight production process. A key planning tool was developed to forecast manpower utilization by mission type and to predict personnel and funding requirements by mission type. In March 1987, the Shuttle Program Office provided the first flight rate projection, targeting four flights in 1988, nine flights in 1989, and a progressive flight rate increase to sixteen flights in 1993.

19

THE TEMPEST

Courage is the price that life exacts for granting peace.

—Amelia Earhart

The Space Station Program was established by President Reagan in his 1984 State of the Union Address. The program was initially envisioned as three small separate orbital facilities: a crew-occupied station, plus two automated platforms for scientific experiments. It was intended to be completed within a decade. There was a complete lack of definition of this work. The only agreed-upon mission was that program would become the largest international engineering project in the history of the world. But once we had built it, what were we going to do with it? No one was focusing on that. In many ways I believed that the Station was the Panama Canal of Space, which would allow participating countries significant technological, scientific, commercial, and political benefits. With the failure of vision to define the space station program, it continued to grow underfunded and uncontained as program participants became aware of the opportunities and benefits.

After leaving Mission Control in 1984, John Aaron and Neil Hutchinson were assigned to form a JSC Level II Space Station office that would develop the integrated design requirements and assign project responsibilities for all four NASA centers. At this time support for the station in Congress was marginal and the scientific community was concerned that funds spent on the station would reduce funds available for other scientific pursuits. Throughout its early history the Space Station Project was in competition with the Shuttle for personnel, financial resources, flight rate, and mission complexity. The best people wanted to work on the Shuttle, not the Space Station. The financial issue became acute with the decision to build a shuttle replacement for *Challenger*.

I normally make the decision to commit MOD resources to work on a program at the point when there is sufficient engineering data that will allow MOD to influence the design. With the grounding of the shuttle for at least twelve months and with a baseline station configuration, I established a MOD office headed by Flight Director Chuck Lewis to support the space station. When the contracts were assigned for Space Station development work, I moved to secure the funding and initiate the design of the Space Station Control Room and training facilities. Considering the Station as a Shuttle payload and the required EVA support during assembly resulted in a needed collaborative agreement between the Shuttle and Station program offices to provide joint funding for the mission support facilities and operations staffing.

JSC Director Cohen, as well as the directors at the Goddard, Marshall, Kennedy, and Lewis's centers all faced the challenge of staffing a program that they believed was underfunded and would probably fail. To gain support for the station program the development responsibilities were split among four NASA centers in work packages. During every year's budget cycle, the station, power, assembly, and user requirements were changed by congressional directives to redesign to reduce costs.

***

On February 3, 1988, Jan Pacek, my administrative assistant, stood by the office door. In muted words she said, "Truly is on the phone." In the early years of the space program, it was not unusual to receive a direct call from Washington to answer a question or receive an action. Knowing Truly well, I picked up the phone and said, "Hi, Dick, what's up?"

His response was sharp, and his words were crisp, "Have you seen this week's *TIME* magazine?" Before I could respond he said, "There is a story there that two of your employees are charging that Mission Operations has compromised the security and integrity of the shuttle flight software." He then continued, "CNN's John Holliman is going to run a story on the Shuttle software allegations tonight and, before running the story, he wanted to hear NASA's side. I want you to talk to him." Holliman came through the news ranks, starting as a sports radio announcer, later becoming a news director of several radio stations. He joined CNN and became the first member of the Washington Bureau.

With only ten months from return to flight (RTF), I immediately recognized the software issue as a major threat to the schedule. I was well versed on the software-processing facility as a result of the work leading up to the consolidation contract, and had confidence in the operations, and safeguards. I developed three pages of notes, then called Holliman. The shuttle flight software technically flies the shuttle and error-free production is demanded in the personnel, process, and software content. Holliman and I talked for an hour. I described the software reconfiguration process, indicated the role that the two Unisys whistleblowers played in the production sequence, and addressed the safeguards and oversight. I concluded by summarizing the critical nature of the software and the "culture" of excellence of the recon team. Closing our discussion, Holliman indicated that he "trusted me" and would not run the story.

When I examined the *TIME* article, I found that nine months earlier two employees had initiated a lawsuit against Rockwell and Unisys for $5.2 million, alleging that their warnings about a compromised safety culture had resulted in workplace retaliation by their employers. Their lawyer sent reports to the NASA Inspector General and to the FBI alleging security breaches, improper software handling, harassment, and telephone threats. During a meeting with the STSOC contract managers, I found they knew of the issue five months prior to the *TIME* article but had not briefed me nor NASA. It was tough to control my anger. The potential for damage to the program was daunting. The *TIME* article suggested that the lessons of the *Challenger* disaster had not been heeded. The article referenced an agency review claiming that, "When NASA rated its Program Managers, 'safety was conspicuous by its absence' in the evaluation and the rush to resume the program had put schedules over safety."[15] The worst damage came from the whistleblowers' claims that any attempt to call out these wrongful practices was met with harassment. That was certainly not the IT! culture of MOD.

To avoid MOD distraction to the ongoing mission preparation, I elected to personally cover the action with Congress and the media. I established a six-person rapid response team composed of three PAO members, the recon section chief, myself, and Jan. I assigned John O'Neill, my Deputy, to lead the Directorate. Tommy Holloway would support the RTF and Jim Shannon would complete the STSOC contractor transition.

Referencing the *TIME* article, Congressman Roe sent a letter to NASA Administrator Fletcher indicating he intended to hold hearings and his committee would provide oversight to the orbiter software issues prior to RTF. We subsequently received a congressional request for the STSOC contract history, flight software history, reconfiguration processes, test performance history, and much more. This request devoured all my time, as well as Jan's and that of the directorate office staff. It also sapped the resources of the JSC Legal Office, literally filling a truck with copies of our records. Most importantly, however, we did not distract the operations for return to flight. "IT!" may have been threatened, but now "IT!" would right the ship.

We began tracking the allegations much as we would normally track mission problems, looking for clear exaggerations and vulnerabilities. Communications with reporters from the *Orlando Sentinel* and the local KPRC space reporter indicated the plaintiffs' lawyer was distributing copies of the highly proprietary contractor performance reports to the media. The book *Crisis Management*, by Steven Fink, was a reference I kept close at hand in my office. I obtained permission to reprint two chapters from the book on dealing with the media: "Controlling the Message" and "Dealing with a Hostile Press."[16]

By distributing the reprints to my staff, I expanded the technical pool available to provide media response when needed. We provided written responses to the hearings of Congressmen Robert Roe and Bill Nelson including a memo I sent to the NASA Administrator, indicating that the allegations had no basis in fact.

On June 6, the *Electronic Engineering Times* interviewed the plaintiffs. The printed story covering two pages was titled, "Shuttle Security Lapses," but it was loaded with technical errors. We prepared a letter to the *Electronic Engineering Times* editor, addressing each of the many allegations and concluded with an invitation to the editor to come to JSC, to meet with the software reconfiguration team, and witness the software processing facility in operation. My PAO team then indicated that the way to get the story to the national media was to publish my *Electronic Engineering Times* letter in the *JSC Space News Roundup*. In the *Roundup* of June 17, 1988, the letter was published with the bold headline, "Kranz Strongly Rejects Trade Journal Charges." The letter was picked up by the national trade and news media.

The *Texas Lawyer* newspaper of September 5, 1988, also questioned the whistleblowers' deal with Rockwell. The lawyer, who had been frozen out of the Rockwell settlement, claimed a congressional staffer, played a highly irregular and unethical role in the whistleblowers' suit by acting as a go-between for Rockwell and the other attorney's client.

Four and one-half months after it started, "IT!" carried the day and I could finally get back to close remaining transition issues and prepare for the Shuttle Return to Flight. The MOD recon team presented me with a "Raising Hell Award" for my letter and I framed it and hung it on my office wall. Responding to a crisis that threatened our entire program, we had applied the "Tough" and "Competent" ethos and succeeded.

On September 29, 1988, STS-26 the Shuttle *Discovery* lifted off, returning the shuttle program to flight status. With Kohrs at the helm in Houston and the STSOC transition behind, I sat once again in Mission Control with the knowledge that "IT!" had carried Mission Operations through adversity and we were well positioned to strongly responded to the new program schedule with a well-trained STSOC leadership team and contractor base. The centralization of all software in the reconfiguration office was paying off as were the highly structured mission planning preparation processes that had been developed in the *Challenger* "downtime." We were ready to fly.

Late in 1988, the Station design was split into two phases to accommodate cost growth. This change provided a single keel, with the crew initially limited to four. Power was reduced and the phase 2 assembly complete configuration was advertised capable of "evolving" to support the Moon and Mars.

20

ROUGH SEAS AND TURBULENT TIMES

Yesterday is not ours to recover, but tomorrow is ours to win or lose.

—Lyndon Johnson

In 1989 Dick Truly received the Collier Trophy while singling out Richard Kohrs and Arnie Aldrich for their effort in returning the Shuttle to flight status. By the end of the year, a formal partnership agreement was made to build the Station as an international undertaking with Canada, Japan, and Europe. I established a Space Station Operations Office to support Cohen and the JSC Projects Office, and initiated the development of interfaces with the Japanese, Canadian, and European offices. During this period, Mr. Akira Kubozono represented the Japanese interests in the Space Station. Although the language barrier prevented direct communications, his sincerity and passion were evident and he and his team were a pleasure to work with on the Station. NASA leadership turned over from Fletcher, to Myers, then to Truly in an acting capacity during April and May 1989. Fletcher, anticipating Station budget cuts, warned NASA the

"Wolf is at the Door." Truly was confirmed as Administrator on July 1, 1989.

In May, I finally made some changes in MOD to address the increasing flight rate, establishing Tommy Holloway as the Assistant Director for the National Space Transportation System (NSTS). Steve Bales was the Assistant Director for Program Support with responsibility for the design, development, and operations of all facilities and the development responsibility for the flight-critical MOD software inventory. This included: software for four orbiters, Mission Control and space station control centers, training systems, and the trajectory and planning systems. MOD now consisted of seven divisions and four major offices. I kept my office small, consisting of a deputy and assistant deputy with secretaries and an executive secretary. The Assistant Directors were collocated with their organizations. Total NASA and contractor manpower were about six thousand at that time and, working with the Shuttle Program and Astronaut Office, we had the plans and data to improve efficiency and significantly reduce manpower in the coming years.

On July 20, 1989, twenty years after the first lunar landing, President George H.W. Bush announced plans for the Space Exploration Initiative (SEI). The Space Station Program, now called "Freedom," was reaffirmed with a plan to return humans to the Moon and years later, Mars, would carry the designation, "A journey into tomorrow." The Station configuration, cost, and schedules remained elusive. However, within the year the goal to return humans to the

Moon was eliminated. The Shuttle Program had recovered well from the trauma of *Challenger*. With a smooth transition in leadership to Aldrich as NSTS Director, we accomplished five flights in 1989. The Shuttle was on track and to a great extent, it was independent of the leadership shuffle at the higher levels of NASA. "IT!" had carried the Shuttle from a disaster through recovery and was now returning to the promise of robust operations.

While the Shuttle was healthy and running smoothly, the Freedom Station was in trouble. The White House, Congress, NASA, and the scientific community wanted to build a Space Station. All agreed it was the next major step in the exploration of space, but they differed on national objectives and significantly on what capabilities a Space Station would provide. Few recognized the magnitude of the task. Congress soon forced another Station redesign to reduce costs in 1990 and Truly transferred Dick Kohrs, an experienced Shuttle Program Manager, to the Freedom Station. I was not surprised with the demanding design evolution and cost increase of the Freedom Station since I had been involved in the initial phases of every American manned space program from the very simple Mercury to the complicated Shuttle. My experience on Skylab led me to believe that the requirements for the Space Station, as a science platform, would be incredibly complex and require an experienced program team. I believed that only those with Skylab experience at MSFC and JSC, had a practical understanding of the requirements and thus the cost for the Station Program.

NASA Administer Richard Truly

Soon after his appointment as NASA Administrator, Truly came into conflict with the Bush Administration and the Vice President Dan Quayle-led Space Council over basic space policies. After Deputy Administrator Thompson's retirement, Truly wanted an insider, like Aaron Cohen, to be selected as deputy. Various groups at Stanford University and the White House opposed building a replacement Shuttle and became critical of the cost and development times for the NASA projects. The call for universities and outside support groups to provide new ideas and technologies further aggravated Truly's relationships with the panel. Outside panels and "expert" recommendations convinced the White House to establish a "new" direction on space and the search for a new administrator began. There were many voices influencing policies and generating actions from the NASA Administrator's Office that were not recognized by the U.S. Congress, NASA Program Managers, nor the Center Directors.

On February 10, 1992, Truly sent his resignation letter to the White House stating that, "The space program is not an endeavor, which as some would have you think, has quick, brilliant, and easy solutions."[17] In a televised message to employees, he indicated he would resign effective April 1. With the deputy position vacant for several months, Truly's planned retirement created a NASA leadership vacuum. Truly, not to mince words, clearly stated when in a time of "Rough Seas and Turbulent Times," there is a Navy saying, "Steady as she goes." On the same day as Truly's retirement announcement, JSC Director Cohen and I dedicated the new Space Station Control Center. After commenting on the thirtieth anniversary of the John Glenn Mission, I predicted that the sound of "Freedom" would resound in these walls for the next thirty years. I was not much of a prophet!

1992 Goddard Memorial Dinner, with Gene Kranz on the right

In March 1992, I received the National Space Club Aeronautic Engineer of the Year Award at the Goddard Memorial Dinner at the Washington Hilton. When the event program began, Daniel Goldin, Dick Truly's replacement as NASA Administrator, was introduced to the attendees. I was seated at a table with several NASA pioneers, astronauts Conrad, Cernan, and Stafford, Aaron Cohen, and Mary Ruth Low, the widow of the highly respected former NASA Administrator George Low. After the dinner, we stood at our table for pictures, and I was surprised that the new Administrator Goldin did not come to visit with the NASA leadership.

When Goldin was selected as Administrator on April 1, 1992, the Freedom Space Station was in serious trouble financially. To reduce costs, further design tradeoffs reduced power, assembly flights, EVA requirements, and the first launch was slipped to 1994. The science users, as well as Congress, were unhappy with the redesign tradeoffs.

In December 1992, in the eighth mission of the year, STS-53 successfully completed the tenth and final mission supporting the DOD. The 1993 manifest consisted of seven missions including another TDRS deployment, two Spacelab missions, and a Hubble Telescope servicing to recover the instrument vision.

"IT!" starts with leadership at the top. In 1992, NASA top-level leadership had been in transition for several years. With Truly's direction and with Aldrich and Kohrs leading the program, the Shuttle had recovered from *Challenger* because it had a stable core workforce and program-level leadership that had worked together for over a decade.

A critical mass is the smallest amount of fossil material needed for a nuclear reaction. The same is true for organizations to function. Leadership must exist, goals must be readily apparent, knowledgeable participants must be present to represent critical interests, and debate must be fostered within the group to reach the objective in a timely and integrated fashion. The space station suffered through a constant change in leadership, poor program definition, inadequate initial funding, internal and external "players" driving objectives, insufficient skilled personnel, and lack of integration. Administrator Goldin faced many challenges, but first he needed to survey NASA, select an experienced staff, and above all, recognize the limits of change. NASA and the aerospace industry itself were aging institutions also in the process of change. The Shuttle was the apogee of engineering achievement that was quickly followed by a significant senior staff retirement and the transition to a new generation of program and operational leadership.

Integration is the most difficult element in program management. It is there that the responsibilities of roles and missions are assigned, policies are defined, budgets and schedules established, and contracts negotiated.

In my years, I had participated in many Level II sessions where the program scope, policies, definition of requirements, assignment of responsibilities, and the control of cost and schedule implementation were addressed. The key to success at Level II is an experienced Program Management Team and unfortunately, Goldin was inheriting a very thin team. He then made poor staff selections and his approach destroyed the motivation needed to build and work as a team. When Goldin repeatedly classified the NASA workforce as pale, male, and stale, he surrendered the ability to build the needed culture and teamwork for success. There is a direct correlation between the character of the leader and the organization achieving its goals.

Dick Kohrs, the Freedom Program Manager

A leader must unify before he can lead, and Goldin did neither. Senator Barbara Mikulski was an ardent supporter of the Space Station, as well as of Dick Kohrs, the Freedom Program Manager. She sent her staffers to Houston three days prior to Christmas 1992, for a briefing on the synergism between the Space Station and Shuttle operations and facilities. We briefed the team on the results of the eighteen-month Operations Phase Assessment Team (OPAT) study that I had conducted with all NASA centers.

When the meeting concluded, Mikulski's staffers indicated satisfaction with our proposed costs, staffing, and operations plans. In a thirty-minute briefing, Mikulski's staff proceeded to provide clear direction to myself on the Station Program challenges that Mikulski expected in 1993, forecasting a stormy year ahead in the most explicit detail I ever heard from a congressional staff.

Kohrs is one of the few NASA leaders we trust. Congress has no sympathy for the NASA management problems and further Station Program disarray will not be tolerated. Few of the NASA bosses are sticking their necks out and they are not challenging the emerging Goldin policies. Lay low in the first one hundred days of the Clinton Administration. Leakers will not be tolerated. Someone is providing one to three faxes per day to the media and the incoming administration. Play it straight and do not be afraid to respond and correct the media.

It was clear to me that the Senator had little respect for Goldin's leadership and felt that his staff was incapable of managing the program and was a prime contributor to the disarray. Dick Kohrs closed the meeting stating, "I need all the help you can give. I am pressing Cohen as far as I can for his help. I can work now only because of the credibility of the MOD team." His closing statement was clear, with Truly gone, now Kohrs, MOD and I all had bullseyes on our backs.

At the beginning of the year, I had a gut feeling that the leadership work ahead was loaded with risk for those in the line of fire. I knew that I had to assume the role as "point man" in addressing operations with the new Headquarters team. On January 6, 1993, I assigned my Deputy, John O'Neill, to act as the Director of all aspects of Shuttle Mission Operations. I considered John as first in the line of succession for my job and, placing him in charge, provided a good opportunity to assess his ability to lead MOD. I then relocated for three months to an offsite facility in Clear Lake, serving as O'Neill's Assistant Director for the Space Station to build the team needed for the coming battle.

My station organization was about the size of large branch, about forty, but had the assigned division level mandate to build the team to support the Station. At that time, I did not know how difficult the coming year would become. I was fortunate that my operations personnel had "IT!" and thrived in crises, met challenges, and as a team under pressure grew in strength. Many other NASA organizations and their leaders, and there were many good ones, at this point worked for months without a clear sense of direction and their morale suffered, personal commitment diminished and, they became victims to Goldin's leadership style. For my part, I lived and enjoyed working once again at the division chief level among the ranks of the controllers. For the first time in many years, I could walk down the aisle, visit with the controllers, and be a direct daily part of the action.

My new working environment was similar to the early STG, a small group responsible for building the space future. I was lucky, however, because every employee in the division had flown Skylab or the Shuttle and we had a wealth of experience at my fingertips in training, space systems, and EVA. We had "IT!" I reported in at the morning telecon like the rest of the MOD divisions, addressing assigned actions, assessing station systems and our progress in developing the Mission Control and training system requirements. I also reported the division complaints upward to "Directorate Chief O'Neill" telling him we needed more parking spaces, building security, and faster turnaround on personnel actions. I had the joy of watching an organization grow, and developing young, new hires into operators. This also provided O'Neill the opportunity to experience the issues in leading MOD and assess how the division elements responded to his direction.

NASA Administrator Daniel Goldin

After resuming my role as MOD Director in April 1993, I had my first meeting with Goldin, thirteen months after his appointment. At Kohrs request, I presented the results of OPAT, the eighteen-month study on Space Station operations plans and costs. I had already provided this material to the congressional staff in December. I am an experienced briefer, but I never got beyond the introduction. After frequent interruptions, Goldin began to lecture the team of senior NASA managers, launch directors, engineers, and astronauts, on his "prior life" and program experience controlling engineering design and costs. Debriefing the meeting, we believed that his sole purpose was to intimidate the attendees and establish that he was in charge

An August 1993 *Los Angeles Times* article characterized Goldin as a "change agent" brought in to reform NASA by purging old managers who voiced reservations. He could replace them with younger, less-opinionated astronauts more willing to toe the new party line.

On August 9, 1993, *Space News* carried an article, "Bush-Era Space Policy Makers Retain Influence." The article spoke of a vision of space not found in any official U.S. government policy, but spoke of a redirected space program, a pending dramatic NASA reorganization, and cooperation with Russia. After reading the material I recognized that we were dealing with two Station Program Offices. The Congressionally approved Freedom Station directed by Kohrs, and a new Space Station Office in Washington established by Goldin and led by astronaut Bill Shepard in Crystal City.

Bill Shepard, a Navy Captain who served with several SEAL teams, came aboard as an astronaut in 1984. He served as a mission specialist on three missions before being assigned to headquarters on Goldin's staff. He was assigned as the Space Station Program Manager in 1993 and instituted actions with the NASA centers to assess design options and costs of several Freedom Station configuration options. He quickly introduced the consideration of "Russian" involvement in the re-baseline.

At the working level, the resulting ever-changing Freedom baseline, who we should respond to, combined with new congressional staffers questions, budget re-justifications, and radical changes in the program content created chaos. Boeing was brought in by a letter contract to support Shepard's Crystal City staff to evaluate further design options that provided "cost savings" by incorporating the Russian elements. The NASA centers, under direction from two different Program Managers, began marching in place and wasting resources before receiving new commands to march off in new directions evaluating options submitted by Boeing.

Hans Mark, the former Deputy Administrator of NASA, had retired in 1984 to serve as the Chancellor of the University of Texas System. I developed respect for Mark when serving as the Mission Operations Director during the early shuttle years. During his tenure as Deputy Administrator, he documented the trials and tribulations and birth of the space station in his book, *The Space Station, A Personal Journey*. Hans and I resumed frequent correspondence after the *Challenger* accident and in 1993, after a discussion on the NASA disarray, he provided some correspondence on the politics behind the chaos. The material was interesting, as much of it came from outside groups planning to take advantage of the *Challenger* accident, using it as an opportunity to move space policy in a new direction. One document was a May 1993 fax on Stanford University letterhead to Ashton Carter and William Perry, stating that NASA and the space industry had ignored President Clinton's call for cooperation with Russia in space. The fax was titled "Outline for a New U.S Space Program," centering on implementation of the 1991 Space Exploration Initiative (SEI) submitted to Vice President

Quayle's Space Council. The ten-page memo stated that "Stanford University," being academic and immune to direct censorship by NASA, was working with various institutions to implement the Space Exploration Initiative.[18]

Russian systems and capabilities were common assets used by the external groups to achieve their objectives. The Synthesis Group recommended phasing out the Shuttle, using international launch systems, and engaging with Russia among other actions. The memo indicated that the NASA staff was suppressing work with internationals to keep the Shuttle and space programs on course. The major target was phasing out the Shuttle and moving to launch on cheaper international systems.[19]

I reread Michael Lawler's *Space News* commentary, "The vision of space at the turn of the century is not found in any official U.S. Policy statement or blue-ribbon report. . . . There is a space policy cabal at the center of changes in the long-term plans for the space program."[20] I was in the middle of this melee and reading the memos, I thought of the Bible statement, "A house divided cannot stand" and I thought of the opportunities lost, careers decimated, resources wasted, and the time when we should have been moving forward in space. That was time that could not be recovered.

When character is lost, all is lost. I realized that working with the Goldin team, I had been naïve. Lacking the necessary political experience, I had trusted them. Kraft's final words to me about *staying strong in my position* had never felt truer. From now on, as long as I continued my journey in NASA, those words would provide the guide for my every action. The "IT!" team chemistry still existed in MOD but now, throughout NASA, it was being threatened by policies driven by individuals striving for positions of power.

21

DUTY

Death is as light as a feather . . . duty as hard as lead,
a life lived in devotion to duty is a hard life . . . but it
is the life of a man.
　　　　　　—*quoted in a letter from Margaret Kranz*
　　　　　　　　　　　　　　　　(Gene's mother)

Character counts, for it is the destiny component in leadership. A true leader must have an untouchable moral authority that wins personnel to their cause. The authority is coupled with a visible and uncompromising personal belief that an individual has been called to accomplish something greater than they would have been able to accomplish by themselves, and that they need a team.

With the appointment of Astronaut Bryan O'Conner as director of the Space Station redesign, I believed he would restore much needed unity to address the station issues. I liked O'Conner. He was a Marine aviator, test pilot, and CapCom for several early missions. He was one of the key links to the *Challenger* investigation and had flown twice on the Shuttle. In 1992, he led a team to establish early planning for the Shuttle/MIR program.

With Kohrs's retirement and that of several of his staff, NASA lost a skilled program management team and friends in Congress. "IT!" was no longer possible for Goldin's Freedom team. "IT!" begins at the individual then spreads through values of trust, respect, and caring and it takes shared experiences to grow. Those Goldin selected as replacements had limited space program leadership experience and many, including the Boeing contractors, were new to manned spaceflight. They reminded me of the "Mercury hired hands," well-intentioned but without the shared task and personal experience. In a span of eight years, the 1982 $17.4 billion Freedom Station budget, now renamed The International Space Station, would grow to $26 billion by 2000.

When President G.W. Bush took office in 2001, NASA revealed further cost growth raising the Station total cost to over $30 billion, a 72% increase during the tenure of Goldin, the man who preached, "faster, better, cheaper" as a philosophy of leadership. NASA explained this increase by stating that the, ". . . Program Managers had underestimated the cost and complexity of building and operating a space station." That part was true, but what was more important is that they never developed the "IT!" to succeed with a complex, challenging initiative.

I had been "grounded" for three decades and the passion to fly had returned and burned bright. I had been working weekends at the Lone Star Flight Museum in Galveston, restoring WWII aircraft, and serving as a crew chief for their beautifully restored F4U-5N Corsair, the fighter that was flown by the U.S. Navy ace Guy Bordelon in Korea. Vacancies in show teams open infrequently and when it was apparent in July 1993 that an LSFM electrician, who served as the flight engineer/crew chief was retiring, I transitioned to flying airshows with the B-17G "Thunderbird."

The cockpit is a busy place in airshows, the pilot in the left seat flies the aircraft, the co-pilot in the right covers altitude, airspeed, and show sequence. The flight engineer scans the instruments, makes the power adjustments, then rise up in the top turret to verify other show aircraft are not in our flight level. The time in the B-17 brought me close to my former boss, Harry Carroll, and his crew *en route* to a target amid flack and assault by German fighters. I never thought I could fall in love with a bomber, but every flight brought me closer to the courage of those who flew. We would fly ten to twelve shows a year flying up to and returning from Oshkosh, Wisconsin, during the summer. On weekends I would train at the nearby Houston Gulf airport with former Flight Director Pete Frank, a Marine aviator, and a Certified Flight Instructor (CFI), to regain my license to fly.

Throughout this difficult period, I missed Kraft. He was a leader, someone to talk to, and one for whom operations was "in the blood." Often without my asking a question, he would nod in agreement to confirm my direction or provide an alternative I had missed. Kraft trusted his gut and the depth of his operations experience. I was now flying solo, but I knew he was aware of the NASA and JSC leadership problems we were facing and would continue to face in the coming years.

Often late at night or in the early morning, I could feel him in my office reminding me of his final directive *to stay strong in my position.*

Now as I approached the end of my professional life, another highly respected strong man, Hans Mark, who I had worked with during the Beggs era came to me in a time of need. Mark was now Chancellor of the Texas University System, and he was following the trepidations of the NASA programs from afar. As a former insider on the NASA leadership, he began communicating his perspective on the policies being established by Administrator Goldin. Hans Mark, James Beggs, and former Center Directors were aggressive in communicating their concerns to both the Bush and Clinton's administration senior staff.

A letter to Mark from the former Administrator Beggs stated, "His (Goldin's) outbursts are making people reluctant to communicate with him. Even the best people are retreating into foxholes. The key issue is trust and confidence. I continue to believe the only action available to us is a letter to the Vice President signed by a group of the alumni."[21]

Another pronounced that, "The trust and openness with which we once pursued problems is gone. We are no longer fail safe, the perils to our programs are increasing."[22, 23]

Fortunately for NASA, the MOD Shuttle teams were on top of the 1993 manifest: launching every six to seven weeks, deploying an additional TDRS satellite, executing four science missions, and preparing for EVA repair of the complex Hubble optics mission by years' end. Their work made me extremely proud. By mid-summer, the Station redesign objective had been clarified to include Russia and, for the first time, MOD had a clear sense of program direction. With an understanding of the ongoing Space Station work at Crystal City, we established teams to provide direct support to the Crystal city systems operability assessments. For the first time, there was effective communications on the proposed design. Knowledge of the actions and objectives was flowing freely between my personnel and the transition team. The MOD Skylab experience in momentum management, solar-power generation, influence of orbit inclination, and other complexities quickly started to pay off in their work. It was refreshing to see the MOD Space Station personnel's morale surge as they now were a significant part of the redesign effort. Former Flight Director, Jay Greene, loyal to MOD and with a trajectory background, was assigned to the Crystal City transition team to fill a critical skill gap in addressing the systems design orbit trajectory inclination issues. I now had a direct contact providing critical information from Crystal City.

Every morning I conducted a meeting to address Space Station systems issues, performance concerns, safety tradeoffs, and a broad range of EVA related assembly issues. I reviewed the listing with the MOD transition team leads, assigned actions, and prepared recommendations for transmittal to Crystal City. Over several weeks as the listing of issues and recommendations continued to grow and with no apparent action from Crystal City, there were decision backlogs that were affecting schedules and costs. The team leads in Crystal City were either directed to ignore our inputs or did not understand the significance of our recommendations.

In August 1993, Aaron Cohen obtained a Zachary Professor position at Texas A&M and retired from JSC. I think he was plain tired of two decades work in managing the complex engineering development of manned space systems. I would miss Aaron as he was a superb engineer. We had developed a deep, mutual respect for each other during the long days and nights at design reviews and change boards. His job was a lonely one. I spent many hours with him, waiting for the West Coast telecon to come to an end.

Various politicians, aware of a potential change in NASA direction, began defending the interests of their constituents. Senator Mikulski was very skeptical of the redesign work and concerned for her state's Allied Signal interests and Texas Governor Mark White got involved with the NASA Boeing proposals to move or eliminate Mission Control. At the same time, the MOD Space Station personnel became concerned on many of the Boeing design baseline details of the Russian/U.S./International Partner Space Station. The redesign team, inexperienced in spaceflight, seemed incapable of addressing technical issues, particularly those related to the EVA assembly. The HQ configuration options were no more than a series of briefing charts. In a joint meeting with the Astronaut Office, Dave Leestma, the Director of Flight Crew Operations, indicated his astronauts shared the same concerns as mine.

After the meeting, Leestma and I prepared, signed, and transmitted a ten-page memo to the NASA Headquarters Assistant Deputy Administrator, General Pearson, which defined our concerns on design, cost, and schedule. The memo provided a detailed assessment of the Russian/ U.S./International Partners Space Station Design Option. The attachments addressed space systems, performance, assembly, EVA access, and shuttle docking. The cover memo closed with the words, "Failure to address these issues will result in a configuration that is complex to assemble and costly to operate."[24] Our memo represented the experience of three decades in manned spaceflight by my controllers and the astronauts.

Leestma and I were aware of the waning support of the Station Program in Congress and believed the content of the memo was critical to keep the redesign moving forward and to avoid rework and save cost. Paul Weitz, the acting JSC Director, was on travel, but Dave and I felt the memo was critically important and, on August 27, 1993, it was sent to Headquarters. Unfortunately, an unsigned draft of our memo was leaked to *Space News* which published excerpts with the headline, "Officials Fear Russian Deal Premature." *Space News*, seeking a comment, located a NASA official who stated, "Kranz has a long history of making bold recommendations to confront complex technical problems. This kind of Kranzogram is not unknown, but the fact that Leestma signed it has added significance."[25]

The Headquarters response was immediate. Three days after we sent the memo, Weitz had returned and his secretary, Mary Lopez, called my office requesting that I join an immediate meeting with Leestma and Weitz. After a brief discussion of the memo, Weitz stated, "My direction is clear. Headquarters says, 'No More Memos.'" He then added, "If you send any more memos in the future, please send them through my office and address them solely to the AA for Space Flight, General Jerimiah Pearson." Leestma and I concurred, and we considered the issue closed.

We never found the source of the leak. In just a few weeks, this would lead to a much greater problem. Another leak of a report by the Office of the Inspector General confirmed that NASA had transferred about $800 million to Russia beginning with the U.S./Russian Mir period and planned to continue to send resources in later years.

We had support within the HQ Space Station Redesign Team in Crystal City. On September 2, 1993, Dan Swint, a member of the Transition Team Office wrote, "I have a different interpretation of the substance and intent of the (Kranz/ Leestma) memo. I think they would be remiss in their duties had they failed to bring their concerns to your attention. The memo provided straightforward advice on the fundamental requirements the configuration must satisfy. . . . There is a growing concern at the centers that the technical issues they have identified are not being adequately addressed." More issues arrived soon. By September, the Goldin/Shepard/Boeing Space Station rebaseline was in serious trouble. The Station configuration, cost, schedule, operating concept, and development roles were all questioned. The work of the Crystal City team was so discredited, it was described within the centers by Station personnel as a "leper colony."

On September 3, 1993, I was called to a meeting in Washington with Bill Shepard and Bryan O'Conner. I quickly found out Shepard's sole purpose was to determine if I was a "team player." He asked two questions. "Who is going to fly the vehicle in orbit?"

I answered, "The United States." I stated this because I knew there were Boeing cost-saving proposals to have Russia provide support from their MIR facilities.

The second question was, "Who do you work for, the Program or the Institution?" I am a Catholic and, remembering the words from Matthew 22:21 (KJV) "Render therefore unto Caesar the things which are Caesar's; and unto God the things that are God's." I stated with a cocky smile, "I work for the Program, for those things that are Program-related, and I work for the Institution for things Institution-related."

This answer did not satisfy them, and they continued to press the question. That day, I was convinced that Mission Operations and my personnel were the final roadblock to whatever path Headquarters was intending to take. A part of the Shepard discussion had been on the operations costs. They thought that having Russia conduct all or part of operations would make the Station Program viable. Boeing did not have the operations skills and if they were seriously planning to give Russia the operations responsibility, how were they going to sell it to the U.S. Congress and the public? On September 7, the new design was released with Russia as a partner. The number of U.S./Russian missions had been increased and the cooperation expanded which essentially merged the U.S. and Russian space programs.

Many years earlier in my career, Marta had given me a beautiful framed, five-inch, square, gold desk placard of Don Quixote and Sancho Panza mounted on horses. Quixote was an "addled" knight wandering in search of chivalrous adventures and, on occasion, engaged in "tilting at windmills." September was pure hell, and I spent much time late at night in the office or at home with a glass of beer or double bourbon. It was then that I would turn on the hi-fi and play the song, "The Impossible Dream" from the soundtrack of *The Man of La Mancha*. As I did, I would sit and wonder what had happened to the NASA I loved and to which I had committed my life.

Often, I would go over to Mission Control and sit quietly in the viewing room during simulations and missions and listen to the teams perform. The "IT!" of their teamwork, the culture, and their professionalism resonated in their words, raised my spirits, and I thanked God daily that the MOD teams carried a much-needed torch of professionalism. At least someone in NASA was doing the right things right. Leadership observations were in my mind now, and I was seeing a vastly different side of leadership. It was leadership without "IT!" It was leadership driven by ego and power, with no respect for the organizations, people, or the mission. It was leadership that was most definitely incapable of running a space program, one that was centered on self and with the single mission to "purge the old NASA." It was an ugly form of leadership, opportunistic, and sustained only by intimidation and by placing personnel in the ranks willing to do the leader's bidding, whatever the cost and unwilling to take a higher road.

Through my life, I had seen leadership in action. I thought of Kraft and Saylor, Carroll, Cohen, Lunney, Aldrich, Kohrs and Hans Mark. I had been led by honest individuals of high character who I trusted, and I had learned much from them. Lessons about life, challenge, triumph, and tragedy. There were lessons of shared risks that led to brotherhood. Lessons of trust where you were willing to follow one in risk. I had been tested before, but now I felt I was coming to the day when I would have my final exam as a NASA leader.

In Korea, while sitting alert in my Sabre, waiting for the klaxon to scramble, I would often mumble a brief prayer. I would say to God, *More than ever I feel the need of having you close to me. At any moment, I may find myself in combat. However rigorous the task that awaits me, may I have the courage to fulfill my "duty."* I was in a different kind of combat now, but I felt the same sense of danger and it prompted me to call again for that same kind of courage.

Twice I went to Hank Flagg in the JSC Legal Office asking if it was the Administrator's prerogative to direct me to cease writing memos. He told me that if I continued to write memos, it could well lead to my facing "personnel actions."

I countered with, "Hank, look at what happened with *Challenger*. Several managers knew of the solid rocket problem, but no one stepped up. Leestma and I were citing cost and safety issues that are key to the survival of the Station . . . are we supposed to shut up?" Flagg was a good man, but he was not going to stir the pot. He just shrugged, and I left.

In his book, *Trust: The One Thing That Makes or Breaks a Leader*, Les Csorba writes of the collapse of Arthur Andersen, a firm in which leadership was absent when it was needed most. After describing the political scandals, culture coarsening, and bigotry at that firm, Csorba then writes, "What America needs in the years ahead is a kind of 'greatest generation' moral leadership produced from a generation that appears reluctant to assume that mantle. . . . Leaders must be 'people smart' if they are interested in taking their followers to a common destination."[26]

For almost a decade, the NASA Space Station seemed in free fall. The Congressional staffers came to Mission Operations to find what was going on since they were not being provided any information from Crystal City. Once again, I talked to JSC Legal to address whether I should provide materials in my possession to the congressional staffers. The materials from Headquarters would have been helpful to the committee and the question on whether to share those put me in an awkward ethical position. To me this was a conflict in business ethics and the legal response was that I should not get involved. The Boeing operations concept and related statement of work that were being reviewed was, once again, focused on the long discredited "better, faster, cheaper" theme as a direct appeal to Goldin and his followers. However, we were again working well, making a lot of progress with the NASA working levels at Crystal City until someone in Boeing or in Shepard's program office started playing hardball and we were cut off.

I took a break for the STS-51 September 12, 1993, launch. It was a long mission, and I spent most evenings with the controllers or by myself in the viewing room. It was incredible to watch the skills that the EVA crews now possessed in comparison to the risky and primitive operation I conducted on Cernan's Gemini 9 EVA. Maybe due to the long duration and my mood, I felt the brotherhood, the "IT!" I loved my people and the work. Returning to the office midway in the mission, I was met by Astronaut Dave Walker, John O'Neill, and some controllers. They received word from Crystal City that Dan Goldin was going to Moscow to negotiate agreements on the conduct of Space Station operations. They had queried Crystal City and all indications were that no one had prepared Goldin for the meeting.

To address the issue, MOD personnel and the astronauts had prepared a new memo containing guidelines that they believed were key to the conduct of operations for the U.S./ Russian International Space Station. I reviewed the twenty-one pages of their proposed guidelines and found them consistent with our experience from the Shuttle, prior space programs and were typical of international best practices for safety and operations.[27]

I also formed the clear impression that astronauts in Houston were not on the same page as those in Washington. In previous weeks, there had been an onslaught of cost-saving proposals originating in Crystal City assigning operations functions to Russia. I was concerned that Goldin might unknowingly compromise key safety and mission success policies we had learned the hard way in Mission Control. I took time away from following STS-51 and prepared a cover memo for the operations guidelines that we believed Goldin should use when in Moscow.

I knew that my memo would violate the "No More Memos" direction we had received, and I considered Leestma too critical to the future of operations. I would be the only signatory and would send the memo via official channels to headquarters. The September 17, 1993, memo was prepared for the concurrence of acting JSC Director Paul Weitz and addressed to Associate Administrator Jerimiah Pearson. There were no other addressees. It was hand delivered to Weitz office by my secretary, Jan Pacek, and waited for Weitz's concurrence.

Two days later, and before the memo was transmitted to Washington, I received a call from Mary Lopez, Weitz's secretary to come to Paul's office. Mary was extremely agitated. When I entered, she said, "Associate Administrator General Pearson is waiting for you."

I was jolted and knew the memo had not been transmitted. While the memo was on Lopez's desk, there had been a leak. Weitz's office was a busy place. I have no knowledge of who saw the memo and described the contents that brought Pearson to Houston. Weitz was not present when I walked in. Pearson congratulated me on my team's performance for STS-51, the nine-day mission deploying the Lageos, a passive satellite containing hundreds of reflectors to obtain geodynamic measurements. I thanked him for his comment.

Then without another statement and in loud and clear words, he stated, "We said, 'No more memos.'"

I simply responded, "Have you read it yet?"

Handing me the memo, he again firmly stated, "We said, *no more memos,*" then muttered something, unintelligible. It felt like he really wanted to dress me down and then order me to do some pushups. I remained silent, then he left the office, placing the memo on Lopez's desk. I could not believe that Pearson, with his operational background and experience, was serving as just a Headquarters messenger boy. I frankly considered it unprofessional of an officer in his position not to consider the issues in my memo. I felt sorry for Pearson, a Marine General and a Navy Test Pilot with tours in Vietnam and Desert Storm, who accepted the task to fly to Houston just to dress me down.

One of the keys to leadership and of "IT!" is not the significance of, "What you have done." What is important is, "What you are doing now." Whatever Pearson had in the past he no longer had "IT!"

Walking back to my office, I knew this was a different NASA leadership than the one that I had worked with previously. On past occasions, I received calls directly from NASA Headquarters; George Mueller asking for a briefing on installing controller consoles on the Apollo tracking ships, Dale Myers questions on crew training, and Truly on return to flight status, among others. In those days, we had trust and mutual respect. Four days later, I was advised that Maine's Senator Bill Cohen had requested a survey and interviews with JSC managers on the use of Russian assets in the current and future space missions.

On the morning of November 23, 1993, I addressed the Mission Operations Directorate at an All-Hands Meeting. I opened with words on the history of manned space operations, the many times we met the challenge of change, adapted, emerged strong, and excelled individually and as a team. All programs provided challenges but, in all prior programs, we could rely on skilled and trusted leadership, credible technical goals for the missions, effective relationships with the program offices and the NASA centers. When challenged, we always met with respect for our teammates. I knew that this was my last formal All-Hands so I gave them the invocation, the *entreaty*, to maintain a positive orientation so that for all missions in the future MOD would provide the role model, the leadership to be strong to emerge from the chaos.

I closed with my guidance for the future:

Today, there is no element of our Station work that is stable. We, mission operations, are challenged to bring the space station program office at Reston and the Boeing contractor online. We will have to teach them about what we have learned in design, development, test, and operations. You must adapt to a multitude of changes in personnel, organization, roles, missions, and leadership. The operations roles and missions we have performed in the past are not understood by the current station program or its contractor. Every day may be a battle and we must show the program their way and what we can do.

My talk was followed by O'Neill and Shannon addressing teamwork, integration strategy, and contract issues between STSOC and Boeing that needed resolution. Following the meeting, I met with Paul Weitz and Hank Flagg in the Legal Office to obtain some resolution of the main contract issue. Boeing believed that they had *carte blanche* to act in all areas of work related to the Space Station and their work was infringing on the STSOC responsibilities because MOD facilities and the related contracts supported both the Shuttle and Station.

For the final time, on December 2, 1993, I resumed my role in Mission Control as MOD for STS-61 to support Flight Director Milt Heflin during the Hubble mission EVA's. The mission was possibly the most critical and technically difficult of all the Shuttle missions. If successful, it would correct a polishing flaw during the Hubble Space Telescope (HST) mirror manufacture. After berthing of the HST in the shuttle payload bay, the mirror optics were planned to be corrected by the installation of five pairs of corrective mirrors during a series of five EVA's, by two pairs of EVA crewmembers. There is an unwritten mission rule for critical EVA objectives. EVA duration is based upon a timeline derived during training and often the primary objective of an EVA can be accomplished with time to spare. The EVA crew, with the Flight Director's concurrence, may elect to continue with tasks planned for a subsequent EVA. If this unwritten rule became controversial for STS-61, I was prepared to support the Flight Director's decision. During the Hubble repair, I spent every day in Mission Control. The teams in Mission Control had many new faces and three new Flight Directors.

One new face in Mission Control was Randy Brinkley. He was from Headquarters and was often addressed as the Headquarters Hubble Mission Director. I was familiar with the Headquarters-types during Apollo and Skylab. Good people, but most came in with only limited knowledge of the missions. Brinkley was different and he reminded me of some of the MSFC Skylab mission directors who had hands-on experience with the hardware and knew the engineers and operators. Flight Director Milt Heflin worked well with Brinkley because he knew the Hubble systems and engineers, asked good questions, and listened to answers. The conduct of the controllers and the Hubble team and crews during the record five EVAs to repair the telescope made me proud to have been their leader. From the earliest days of space, when I was initially tagged "General Savage" by my controllers, I was committed to a cause defined by simple words, "Discipline, Morale, Tough, and Competent" and I prayed they were embedded in the minds and hearts of my MOD. The mission days during the Hubble repair were an absolute joy.

With the STS-61 landing at 11:25 p.m. CST, December 13, 1993, I stepped down from the Mission Operations Director console and closed out my life in Mission Control. It was much like leaving the cockpit of a Sabre many years before and alone in my office, I shed some tears. The success of the HST repair restored much of the NASA image with the public, Congress, and the ranks of the controllers. It was the final demonstration of "IT!" I would have with my team.

The Christmas holiday period was entirely too short. There was no escaping the swirling rumors nor the internal revolution at the top levels of NASA. The General Accounting Office came down hard on NASA Headquarters, with a listing of fourteen major topics on the Space Station and Russian involvement. The Kranz/Leestma concerns in our memo, were just two of their topics. Other topics included training, proposed trajectory options, Russian commitments, costs, schedules, and several questionable payments to the Russians.

On Wednesday, January 6, 1994, Carolyn Huntoon was named JSC Center Director with George Abbey as Deputy Director, and Paul Weitz as Assistant Director. The following day, I submitted my resignation to Huntoon effective March 3, recommending that John O'Neill should assume my position immediately as I began preparation for retirement.

My talk to the Directorate personnel was short, as I think they knew what was coming. I told them I had submitted my resignation and had expressed my concerns to NASA Headquarters and to Director Huntoon on the poor leadership, lack of competence, and trust at NASA Headquarters. I thanked God that the Shuttle Program was rock solid, and I commended them for their performance:

The future is up to you, all of you. You must hold the fort until things got better and they will. . . . You must continue to believe in your mission, your team, and yourself. You have IT! you are a great team, with great leadership, and you have the capabilities to survive in difficult times. . . . Thank you for your work and the honor you gave me as your leader.

Many people dropped into the office to talk. My final days were spent with Jan Pacek, the first Mission Control secretary and my wingman, boxing up my files and notebooks and closing out my office to end a long, well-loved career. We had worked together for over two decades and witnessed challenge, triumph, and tragedy. We were a critical part of the team chemistry, the "IT!" that carried our Nation forward into space.

I was invited to the Cape on February 3, for the STS-60 launch. Amazing though it may seem, I had worked with the Cape Kennedy launch team for three decades but had never witnessed a launch. The launch team was not about to let me retire without joining them and their team for a launch, and then feasting on the traditional post-launch beans. Many of the Cape Shuttle team members had supported my Apollo launches and it was like a reunion of old friends. I spent four days at KSC, visited the facilities, and the teams I had worked with in the years before.

With my approaching retirement, I was visited by many of those I had worked with or for. Kraft came by and asked me about my future. We talked about "our time" of clear goals, expressly given, of leaders with the trust to tell it straight and accept the consequence of their actions.

The day I left my office, I left a memo for Astronaut Dave Leestma, Director of FCOD, containing testimony by Admiral Rickover on responsibility. It addressed the Submarine Thresher design and outfitting. The Rickover memo described the environment and lack of accountability for the 1963 submarine Thresher loss during a diving test with all crew aboard:

"During the six years of the submarines design, the Portsmouth Naval shipyard had three shipyard commanders, three production officers and five planning officers. The Bureau of Ships during this period had two Chiefs of Bureau, five or six chiefs of the design division and three heads of the

Submarine Type desk. Endeavors succeed or fail because of the people involved."[28]

I closed my letter to Dave with the words, "With the emphasis on reinventing NASA, we must assure that individual responsibility is not forever lost. When the dust finally settles on the trials and tribulations of our programs, we must leave individuals accountable for design, development, and operations to continue our efforts in space."

***

"Brass Rhythm and Reeds" band, conducted by Flight Director Milt Heflin

The smooth sound of the MOD, eighteen-member "Brass Rhythm and Reeds" under the direction of Flight Director Milt Heflin filled the two Lone Star Museum hangars on March 26, 1994, celebrating my retirement. The hangars were filled with many visitors: Harry Carroll; Astronauts Truly, Gibson, Seddon, Kerwin, Haise, Bean; and former NASA Deputy Administrator Hans Mark. My boss Chris Kraft joined a baker's dozen of Flight Directors for pictures and joined the training teams, staff, and neighbors for the night's celebration.

My retirement was a return to an earlier time in my life. I began working at the Lone Star Flight Museum, flying as crew chief on the B-17 Flying Fortress *Thunderbird* at ten to twelve airshows annually. I converted our garage to a white room for construction of a fully aerobatic biplane, the Murphy "Renegade."

The Murphy Renegade- a six year construction project

Astronaut Bob Overmyer followed construction and planned to develop the test plan and perform the flight test. Sadly, Bob died in March 1996 while flight testing the Cirrus VK-30 homebuilt aircraft. The Marine fighter colors of the

1930s provided my Renegade paint scheme. The Renegade was located alongside the Lone Star Museum Grumman F3F on September 13, 2008, when Hurricane Ike came ashore destroying my aircraft.

The 1994 PBS documentary *Apollo 13: To the Edge and Back*, was used by the Public Broadcasting System for fundraising and took me to Washington D.C. While chatting between the program segments with former controllers, a literary agent, George Greenfield, heard our discussions and contacted me to write a book of our adventures. Writing of my controllers had always been a dream and I had eager support for team interviews for the book. Over a three-year period, I drafted the story of the teams of Mission Control in my book, *Failure Is Not an Option.*

Following Jim Lovell's recommendation, I began a speaking career in 1995, initially at small colleges and while passing out LEGO toys, talking to kids at department stores. With the success of the *Apollo 13* movie, my speaking engagements quickly morphed into engagements covering most of North America and Europe, often with astronauts Lovell and Haise. To my great pleasure, airshow speaking provided the opportunity to fly with the performers during the show rehearsals.

***

During the Shuttle *Columbia* launch on January 16, 2003, a piece of foam from the external tank attachment separated and damaged the wing leading edge. During reentry, the breach in the leading edge allowed superheated air, about 5,000 degrees, to enter the wing structure, melting the aluminum spar and progressively destroying the wing. The seven astronauts were killed in the reentry breakup.

As the successor to Dan Goldin, Sean O'Keefe was initially brought in to address a five-billion-dollar Space Station cost overrun but his focus quickly turned as he became enmeshed in the Shuttle *Columbia* accident. After participating in a *Columbia* accident briefing, I was called to Headquarters in November 2003 by O'Keefe. He did not provide a name but said "someone" recommended that he should talk to me about the accident report and the NASA/JSC/ Mission Control chain of command. During our discussion I indicated that I believed the leadership team had lost the "flight test mentality" I had learned from my flight test boss, Harry Carroll, and my early years with Kraft. I discussed the changing culture of JSC stating that JSC from its formation was an engineering- and operations-driven center. Without a succession plan, in the post-Cohen era, JSC center leadership changed four times in the decade after Cohen. I also believed the JSC engineering and operations functions were under stress and had been compromised due to constant change and the lack of strong experienced leadership. Flight Directors now in Program Offices were exercising their former roles in the Mission Management Team, diminishing the Flight Directors' mission decision responsibility. I quoted a second statement in the *Columbia* report, "Changes in organizational structure should be made only with careful consideration of their effect on the system and their possible unintended consequences."

***

In 2004, I was again called to NASA Headquarters by Administrator Sean O'Keefe to speak to NASA employees on the "Constellation Program." This was O'Keefe's response to President Bush's goals in the Vision for Space Exploration, and I toured several NASA Centers in support of the Constellation Program.

***

In 2007, Chris Kraft and I were added by NASA Administrator Mike Griffin to the list of NASA's first generation of explorers and honored as Ambassadors of Exploration. Chris was recognized as the architect of Mission Control while I was recognized for relentless dedication to crew safety and mission success. We joined the eminent list of honorees including astronauts Alan Shepard, U.S. Sen. John Glenn, and Neil Armstrong. The lunar fragment I received from Apollo 17 now resides at Central Catholic High School, Toledo Ohio, where my career began.

***

On May 2, 2008, I received a letter from Senator Daniel K. Inouye. Michael Griffin, the current NASA Administrator had nominated me as an Advisory Board Member of the Coalition for Space Exploration.

On June 20, 2013, astronauts, program managers, controllers, and I met in the Oak Grove at JSC to celebrate the tree planted in memory of Neil Armstrong. That evening I spoke of "Our Time" at the Astronaut Reunion. We were the generation who responded to the challenge to plant America's flag on the Moon. We had grown and lived through three wars and had served and seen our countrymen die for our freedoms. We lived as explorers and charted America's path in space. We knew about high risk and the grief for our friends who gave their lives to the effort.

"We knew the triumph of high achievement, and when we failed, at least we failed while daring greatly. Our place will never be with those cold and timid souls who neither knew victory or defeat."

Jeannie, Lucy, Brigid, Carmen, Joan, Mark
Marta, Gene

While writing this book, I have been called to brief the new NASA Artemis generation recruits. Like Harry Carroll, I am not all used up. I cannot live out my golden years in quiet satisfaction.

FOUNDATIONS OF MISSION CONTROL

- To instill within ourselves these qualities essential for professional excellence:

    Discipline ... Being able to follow as well as lead, knowing we must master ourselves before we can master our task.

    Competence ... There being no substitute for total preparation and complete dedication, for space will not tolerate the careless or indifferent.

    Confidence ... Believing in ourselves as well as others, knowing we must master fear and hesitation before we can succeed.

    Responsibility ... Realizing that it cannot be shifted to others, for it belongs to each of us; we must answer for what we do, or fail to do.

    Toughness ... Taking a stand when we must; to try again, and again, even if it means following a more difficult path.

    Teamwork ... Respecting and utilizing the ability of others, realizing that we work toward a common goal, for success depends on the efforts of all.

- To always be aware that suddenly and unexpectedly we may find ourselves in a role where our performance has ultimate consequences.
- To recognize that the greatest error is not to have tried and failed, but that in trying, we did not give it our best effort.

LEADERS' QUOTES

Dr. Christopher C. KRAFT—Director, NASA Johnson Space Center, Houston, Texas

- Most of us were the lucky ones to be at the right place at the right time. We should not take it for granted. We are privileged to be in our position today.
- You must always try to be the best at what you perform. That is what will keep you happy regardless of the outcome.
- Consider the other person when making decisions that count.
- Be courteous and respectful to both your friends and opponents.
- Make sure you enjoy your work and if you cannot, move on and accept the next challenge.
- It takes a lot of thinking, a lot of compromise, a lot of talking and a lot of doing to learn to be a leader.
- Accept the Challenges placed before you.
- You cannot lead unless you know what a guy has to do to do their job. Understanding the job is paramount to becoming a leader.
- You cannot become emotionally involved because of the danger of deciding for the wrong reason.
- Faith is important to your well-being, and it should be regular in order to maintain discipline in your everyday life.

- Love is God's most precious gift.
- If you are surrounded by people, you have absolute faith, trust, and confidence then you can make difficult decisions.
- A good leader is capable of letting a person do the job their way. That is the only way that they will learn the job.
- Surround yourself with talent and create an environment that lets them do their job.
- Give responsibility, authority and build confidence and trust.
- Let your people know when they do well.
- If you select the wrong person and have given him the authority and responsibility and the job is not getting done, you must get rid of him.
- There will be a lot about a job that you do not like but you have to be willing to do it all.

John HODGE—Chief, Flight Control Division, NASA Johnson Space Center

- Develop your own vision and means of expression, trust yourself to do right, seek a good mentor and listen to your inner voice.
- Trust is the underlying issue in getting people on your side.
- There is no leadership if shared values have disintegrated.
- Trust is pivotal in a leader's success.

- Person to person is key, followers feel their leader understands. them and has a feeling for their capabilities.
- The team forms when the leader helps followers to develop their own initiative and use their own judgment.
- People must know how they are doing.
- Six Phrases for daily use:
  What is your opinion?
  You did a good job!
  I made a mistake.
  Will you. . . .
  Thank You!
  We. . . .
- Look for simple answers but question everything.

Morris "Jack" COLEMAN—Primary Flight Instructor, Spence Air Base, Moultrie Georgia

- Leaders learn from leaders. It is the company you keep. If you associate with leaders, you may become one.
- Part of the role of a leader is to teach the next generation and provide the example.
- The characteristics that allow a person to be receptive to the lessons of leaders is the groundwork of the parents, teachers, and friends.
- Think three-dimensionally. Visualization must be second nature, flying the airplane must be in the

background, it must be automatic so that it consumes only a fraction of your mind, you must tune every sense so it balances all the requirements in a split second so that you can do other things with a total focus to accomplish your mission.

- Leaders recognize the need for decisions; thus, they plan ahead and know how to use the Clock.
- Every flight is Go/No Go. Your takeoff, air work, and landing are all Go/No Go. Each of these requires you to use your judgment on whether it is safe to continue or take an alternate. Your judgment is your key to longevity and to control the risks of flight.
- The landing is not over until you have shut down the engine, the chocks are in place, aircraft tied down, and forms filled out.
- If the student says, "I'm not ready when I say he is ready" then I say, "That is all I need to know and don't come back to me." If you can't go when I say you are ready to go, then you can't go when you tell me you are ready.
- Student/Instructor personality conflicts are real so try to make changes when you recognize it is hard to impart knowledge or give instruction.
- The early recognition of a washout is when a student does not transfer what you are imparting to him, and you can't get him to react.
- Total immersion is the way to develop skills.

- The criterion for a solo is that the student must have confidence in themselves, enough skill to make a good landing and judgment to recognize a bad landing and initiate a go-around.

Ralph SAYLOR—Chief XGAM 72 Flight Test McDonnell Aircraft Holloman AFB, New Mexico
- Respect those who have dirty hands.
- Dealing with people you do not control is easy if you cultivate them as friends and ask their help.
- Respect for authority.
- Personal Creed—Know right from wrong, what you *stand for.*
- Clear objectives make the difference.
- Leaders have a positive identifier.
- Teaching never gave a direct order—defined the objective/outcome.
- You need balance in your life. Importance of leisure time activities.

Harry CARROLL—Chief Flight Test Data Reduction, McDonnell Aircraft
- It's easy to lead when everything is going smoothly. . . Leaders earn their pay when the going gets tough.
- While there are several characteristics that generate and sustain trust there are four which dominate:

    Integrity—Leaders are whole, they honor their commitments and live by strong values.

Reliability—Leaders are there when it counts, and they are there for the team.

Constancy—Leader's performance is stable; they do not create uncertainty.

Congruity—Leaders walk the talk. There is no gap between what they say and do.

- The team wants three things: direction, trust, and hope for success.

Col Francis GABRESKI—excerpted from *Gabby: A Fighter Pilot's Life*, Francis Gabreski and Carl Molesworth, Bantam Doubleday Dell (1991)

- "I was friendly but not familiar with my new pilots. I put on a stern face. I wanted them to know from the word GO that this was serious business they were getting into, and their squadron commander was going to demand top performance."
- "The Warrior culture is the key to mission accomplishment in critical and high-risk times."
- "One of the biggest jobs as Wing Commander was to keep the men motivated under adverse conditions. I spent a lot of time out on the flight line talking with them and looking to see what they were doing. I had a special red Jeep that I drove, and part of the reason for the paint job was so that the men would be able to tell I was out there with them, even if I didn't happen to stop by each revetment every day."

- "In Operations fear lasts only as long as the anticipation, and then the training takes over. You do what you have to do, destroy him before he destroys you."
- "I knew I was ready as I could be . . . but I still felt the dread of the unknown. I'd come too far, and I was not going to disgrace myself, my family, or my country."
- "I worked hard and listened closely to the instruction we were getting. I think that this probably served me better in the long run than if I'd been one of the top students. Then I'd have gotten cocky and wouldn't have paid so much attention."

Eugene F KRANZ—Director NASA Mission Operations

- When in doubt, and nothing makes sense trust your gut.
- Know who you are, your strengths and weaknesses.
- Above all know whether you are driven to do something or be something. Those who are driven to do something will persevere in the long haul.
- Leadership, trust, and shared values provide the foundation of a team.
- If you live by what you stand for you will see it reflected by your team.
- Do not be a stranger to yourself.

- Culture is defined by the values shared by a team. Culture creates a chemistry that is a force amplifier which amplifies the individual as well as team talents.
- Train your team in the five communications C's learned as a Forward Air Controller: Clear, Correct, Concise, Convincing, and Crisp.
- The Team is the product of Leadership, Trust, and Shared Values.
- Mentally and emotionally get in the Zone before the start of the day—Establish oversight, use a Wingman.
- Leaders are expected to be out right from the start.
- Leaders are visible—Vest, Gabreski, Saylor, Carroll.
- Be predictable, constant, and personally set the frame of reference for performance.
- Leaders are approachable—open door, telecoms.
- Leaders are emotional and have humor.
- Checklists provide structure.
- Leaders know their team by name and those who support them. Security, Personnel, Resources, etc.
- Leaders take the same medicine as the staff and team.
- Healthy tension between team members and leaders is good.
- Proper disrespect.
- When you attack, go for the kill.
- A successful meeting leaves nothing to chance.
- Plan, develop alternatives, train, and execute.
- Wearing two hats is not tough if you use the same structure, same processes, and same team.

- Be human—Show emotion but never let them see you afraid.
- There is no such thing as a, "Small personnel problem."

KRANZ—Impediments to Leadership
- Ego
- Limited personal values
- Wrong leadership model—imitates others
- Self versus mission orientation
- Lack of commitment
- Not comfortable with people
- Poor listening and observation skills
- Unwilling to experiment
- Unwilling to risk failing
- Experience is not a teacher

GLOSSARY -TOUGH AND COMPETENT

56M	1956 Pilot Training Class
Abort	Time Critical event termination
ACM	Air Combat Maneuvering (Dogfighting)
ADF	Automatic Direction Finder
AFB	Air Force Base
AM	Airlock Module
AOD	Aircraft Operations Division
ASTP	Apollo Soyuz Test Project
ATDA	Agena Target Docking Adapter
ATM	Apollo Telescope Mount
Avro	A.V. Roe and Company
CapCom	Capsule Communicator
CCB	Program Change Control Board
CF-100	Canadian jet trainer
CF-105	Canadian Supersonic Interceptor
CMG	Control Moment Gyroscopes providing Skylab attitude control
Cryo	Oxygen and Hydrogen gases stored below -150deg C.
CSM	Command and Service Module
CSQ	Coastal Sentry Quebec tracking ship
DMZ	Demilitarized Zone
DOD	Department of Defense
DOI	Descent Orbit Injection maneuver

EECOM	Electrical, Environmental, and Communications Engineer
ETOPS	Extended Range Twin Operations – Commercial Aircraft
FAC	Forward Air Controller
FBS	Fighter Bomber Squadron
FCD	Flight Control Division- provides majority of MCC controllers.
FCOB	Flight Control Operations Branch- Provides Procedures, AFD, develops Mission Rules
FCOD	Flight Crew Operations Division
FOD	Flight Operations Division
Foundations	Foundations of Mission Control changed to Mission Operations
FRR	Flight Readiness Review
FWS	Fighter Weapons School
G	Acceleration force in relation to Earth gravity
GCA	Ground Controlled Approach
GMT	Greenwich Mean Time, often referred as "Z" Zulu time
Go No/Go	Decision to proceed or terminate
GSFC	Goddard Spaceflight Center -Greenbelt Maryland
GT-(X)	Gemini Titan (mission)
HQ	Headquarters
HST	Hubble Space Telescope.
Hun	F-100 nickname

IT!	Task and Social Chemistry
JOC	Joint Operations Command (Japan)
JSC	Johnson Space Center
K-55	Korean Airfield at Osan Ni
King 1	Holloman Range Control Center (N.M)
KSC	Kennedy Space Center, Florida
L- (days)	Days remaining to launch day
LACIE	Large Area Crop Inventory Experiment
LBO	Light Beam Oscillograph
LM	Lunar Module
LRC	Langley Research Center, Langley Field, Virginia
LSFM	Lone Star Flight Museum
MA-(X)	Mercury Atlas- Flight Number
MAC	McDonnell Aircraft Company
Mach(X)	Percent Speed of Sound
MaxQ	Maximum Structural Dynamic Pressure point,
MCC	Mercury Control Center, later Mission Control Center
MET	Mission Elapsed Time since liftoff
MDA	Multiple docking Adapter
MIT	Massachusetts Institute of Technology
MOCR	Mission Operations Control Room
MOD	Mission Operations Directorate or Mission Operations Director
MPAD	Mission Planning and Analysis Division
MR-(X)	Mercury Redstone (X) mission

MS	Protective Micrometeroid and Thermal caver
MSFC	Marshall Space Flight Center- Huntsville, Alabama
NACA	National Advisory Committee for Aeronautics
NAR	North American Rockwell
NASA	National Aeronautics and Space Administration
OWS	Orbital Workshop
PAC	Pacific Airmotive Company
PC+2	Maneuver for Apollo 13 return
PDI	Powered Descent Initiation maneuver
pH	Measure of acidity or base of a liquid
REFSMAT	Reference Matrix
RKV	Rose Knot Victor tracking ship
ROC	Republic of China
ROE	Rules of Engagement
ROTC	Reserve Officers Training Corps
RTF	Return To Flight
SEI	Space Exploration Initiative, established by President Bush
SEVA	Stand Up EVA
STS-(X)	Shuttle numerical launches
SimSup	Simulation Supervisor
Skylab	United States First Space Station
SOCAR	Systems Operations Compatibility Assessment Review

SPAN	Spacecraft Analysis Team
SRB	Solid Rocket Booster
SSR	Support Staff Rooms
STA	Shuttle Training Aircraft
Stay/NoStay	Decision process after lunar landing
STG	Space Task Group
STS-(X)	Space Transportation System (Flight)
STSOC	Space Transportation System Operations Contract
TPCO	TelePrinter Control Operator
T-zero	Time to rocket Ignition
XGAM	Quail Guided Air Missile (Experimental)

MISSION CONTROLLER POSITIONS

AFD	Assistant Flight Director
Agena	Gemini Target Vehicle controller
Booster	Booster Systems Engineers (2Gemini and 3 Apollo)
EECOM	Electrical, Environmental, Cryogenic, Structure, Fuel Cells
EGIL	Electrical, General Instrumentation and Life support
EVA	Extra Vehicular Controller (2)
FAO	Flight Activities Officer
FLIGHT	Flight Director- Team Chief
GNC	CSM- Guidance, Navigation Systems, Propulsion, Attitude Control and Computer Hardware

GUIDANCE	Call sign GUIDO - CSM and LM Specialist in Navigation and Computer Software. During Gemini included Titan II Guidance Systems
CONTROL	LM- Guidance, Navigation Systems, Propulsion, Attitude Control and Computer Hardware
INCO	Communications management for all airborne systems, voice, telemetry, tracking, video, command and recorders
MOD	Mission Operations Director
Network	Network controller
PAO	Public Affairs Officer
Procedures	Procedures controller
RETRO	Retrofire Officer
Surgeon	Flight Surgeon
TELMU	LM Telemetry, Electrical, Life Support, EVA

ACKNOWLEDGMENTS

The entrance walkway of Space Center Houston is comprised of six quadrants of engraved bricks: Earth, Moon, Mars, Jupiter, Venus, and Mercury. When I completed my first book, *Failure Is Not an Option*, I selected fifteen bricks within the Moon quadrant. I had them engraved with names of my "Teachers" and their critical "lessons" that prepared me for the early years of a successful life journey.

My wingman during the early years of Mercury and Gemini was Joe Roach, an Air Force Captain in the Alaskan Air Command. Hal Miller taught contract management to a novice branch chief.

Arthur Ashe wrote, "Success is a journey, not a destination. The doing is often more important than the outcome." *Tough and Competent* relates the continuation of my journey and the lessons from the teachers who expanded my abilities beyond the landscape of Mission Control including: Aaron Cohen, Kenneth Kleinknecht, Dick Kohrs, Dick Truly, Hans Mark, and Bill Tindall.

The Space Shuttle program opened the sphere of spaceflight to commercial, scientific, and military users requiring a human spaceflight system not only capable of payload delivery but also payload and cargo return. This dynamic, high flight-rate reusable system required extensive training and orientation for both operators and users. Mike Hawes was a key member of a group of Notre Dame graduates that developed payload support capabilities and whose Saint Patrick's Day celebration added a new social tradition to "IT!"

Shuttle operations demanded a new range of contractor management skills. John O'Neill, an experienced Apollo mission planner, was my Deputy and my wingman. Jim Shannon was critical in the consolidating of sixteen contracts into a single contract with two subcontracts involving 5,000 personnel as well as assessing contractor performance. Cecil Dorsey covered the loose ends during the MOD/JSC organizational transition. In this dynamic new era of spaceflight, a long list of controllers assumed leadership roles to assure crew safety and mission success: Tommy Holloway, Jerry Bostick, Randy Stone, John Harpold, Bob Holkan, Carl Shelley, Steve Bales, Jack Knight, Frank Hughes, Ed Fendell, Chuck Lewis, and Earl Thompson.

For those I have overlooked or didn't specifically acknowledge, you all played invaluable roles. Chalk it up to my advancing age but know I remain thankful.

George Greenfield, my literary agent for *Failure Is Not an Option* encouraged me to continue writing. He introduced me to Todd Brewster who assisted with the continued writing of this book and developing a book proposal. I enjoyed working with Todd and learned much from our time together. Although our book did not get off the ground, Todd convinced me to continue because he believed in the story.

I experienced "team chemistry" many times in my former work and I began an enjoyable rewrite of the book using observations obtained from Joan Ryan's book, *Intangibles*. Dutch von Ehrenfried, a key member of my early space team, has self-published eleven books on space and technology. He provided invaluable advice to get me started on self-publishing.

Support from NASA and the Johnson Space Center for a book on the history of organizations and events from a half a century ago was a major challenge. With help from Julie Helton and Mary Wilkerson, the details of JSC leadership and program management were derived from phone books and other materials. Jane Buckley of the JSC Photo Operations Group researched and provided many of the photos.

Alpha Flight Director, Bill Reeves, President of the Manned Spaceflight Operations Association, provided controller contact information and his continuing work sustains the Team Chemistry of Mission Operations.

Working with Gatekeeper Press is reminiscent of working with my mission operations teams. Sarah Duckworth has been a great Flight Director and the support of her team has been superb. Jessica Bushore was spectacular in finding the "hidden leadership lessons" among my words. Cover Design and Formatting was done by Davor Dramikanin and Nivash Prabakaran was the eBook converter.

The most important of my teachers has been my wife Marta, who over six decades provided the foundation of my life and that of our six children. We are fortunate that our family is very close and exhibits the very best of team chemistry, the "IT!" Our oldest, Carmen, holds a BS in nursing and an MBA. She previously owned and managed home health and hospice organizations and recently returned to "hands on" cardiac critical care. The remaining five had careers in the space program. Lucy served as a program planning and business manager for the Space Station, Shuttle, and Orion Programs. Joan served in the mission control computer complex, in crew training and is currently working for Axiom Space fabricating soft goods for space missions. Mark worked in the Shuttle Avionics and Integration Lab (SAIL) and is now a safety supervisor in the oil industry. Brigid has been a financial analyst for Oracle and Boeing. For over a decade, Jeannie led external communications for United Space Alliance; served ten years as a senior space policy advisor for the Chairman and Ranking Members of the U.S. House, Space and Aeronautics Subcommittee; and continues working and continues her work in the Space Program. Jeannie conceptualized the book cover design and provided oversight throughout the writing and development.

Eugene Francis Kranz

Gene Kranz, a legendary NASA Flight Director, is best recognized as the leader of the flight controllers "Tiger Team," who brought the Apollo 13 spaceship safely back to Earth on April 17, 1970. Originally from Toledo Ohio, Mr. Kranz graduated from Parks College of Saint Louis University in 1954 with a Bachelor of Science in Aeronautical Engineering, and served in the U.S.  Air Force as a fighter pilot and a flight test engineer before joining NASA in 1960. He assumed Flight Director duties for the Project Gemini and Apollo missions, leading the controller team for America's first lunar landing. In 1983, Mr. Kranz was assigned as the NASA Director of Mission Operations with a workforce of over 5,000. After serving in Mission Control for over 100 launches, Kranz retired in 1994, turning to motivational speaking and writing. His book on the early manned space program, *Failure Is Not an Option*, was a *New York Times* best seller and was selected by the History Channel for a documentary on Mission Control. He has received numerous awards, including the National Space Trophy, the Presidential Medal of Freedom, and is enshrined in the National Aviation Hall of Fame. A Texan for over five decades, Mr. Kranz and his wife, Marta are the proud parents of six children.

ENDNOTES

1　Charles Murray and Catherine Bly Cox, *Apollo: The Race to the Moon* (New York: Simon and Schuster, 1989) 283.

2　Archibald Henderson Stewart, *George Bernard Shaw: His Life and Works* (Cincinnati: Kidd and Company, 1911).

3　Robert Coran, *Boyd:The Fighter Pilot Who Changed the Art of War* (Little, Brown and Company) 72.

4　John Schultz, *Songs from a Distant Cockpit,* 2nd ed. (John J Schultz, 2013) 75.

5　Ernest Hemingway, "London Fights the Robots, *Colliers Magazine,* (August 19, 1944).

6　General Michael Moseley Air Force Chief of Staff, *The Airman's Creed* (April 18, 2007).

7　Henry Dethloff, *Suddenly Tomorrow Came...: A History of the Johnson Space Center*, (National Aeronautics and Space Administration, 1993) 22.

8　Chris Gainor, *Arrows to the Moon*, (Apogee Books, 1971) 32.

9　John Gardner, *On Leadership*, (New York: The Free Press, 1990) ix.

10　Malcom Johnson, Tindallgrams, (Charles Stark Draper Lab Apollo: 1996).

11　Gerald D. Griffin, "NASA Johnson Space Center Oral History Project," NASA, March 12, 1999, https://historycollection.jsc.nasa.gov/JSCHistoryPortal/history/oral_histories/GriffinGD/griffingd.htm.

12 John W. Aaron, "NASA Johnson Space Center Oral History Project," NASA, January 26, 2000, https://historycollection.jsc.nasa.gov/JSCHistoryPortal/history/oral_histories/AaronJW/aaronjw.htm.

13 Joan Ryan, *Intangibles: Unlocking the Science and Soul of Team Chemistry*, (New York: Little Brown and Company, 2020), 240.

14 Dethloff, *Suddenly Tomorrow Came.*

15 Ed Magnuson, "Putting Schedule over Safety," *TIME Magazine*, Feb. 01, 1988.

16 Steven Fink, *Crisis Management: Planning for the Inevitable*, (American Management Association: 1986) 92 and 103.

17 Andrew Lawler, "Truly Ouster Was Two Months in the Making," *Space News*, February, 1992.

18 Bruce Lusignan, *FAX Communications Satelllite Planning Center*, (Stanford University: May 20, 1993).

19 Lusignan, Fax.

20 Andrew Lawler, "Bush Era Space Policy Makers Retain Influence," *Space News*, 8/9-15.

21 James Beggs, Letter to Hans Mark (January 7, 1994).

22 John McElroy, Letter on UTA Arlington letterhead to Hans Mark (December 1, 1993).

23 Hans Mark, Memo to Honorable John Gibbons (February 1, 1994).

24 Director of Mission Operations (Kranz) and Director of Flight Crew Operations (Leestma) to NASA Headquarters Assistant Deputy Administrator, "Results of Preliminary Assessment of the Russia/U.S./

International Partners Space Station Design," (August 27, 1993).

25   Andrew Lawler, "Space News Officials Fear Russian Deal Premature," *Space News.*

26   Les Csorba, *Trust: The One Thing that Makes or Breaks a Leader*, (Nashville: Thomas Nelson, 2004) 113.

27   Director of Mission Operations (Kranz) to NASA Headquarters Associate Administrator for Space Flight, "Operations Principles for the U.S./Russian/International Partner Space Station," (September 17, 1993).

28   Admiral Hyman Rickover, *Statement on Responsibility during the Submarine Thresher Hearings.*

INDEX